"When I first encountered the patients whom I later wrote about in *Awakenings*—patients who had all had the epidemic sleeping sickness, encephalitis lethargica, forty or more years before, I could find no good general account of the epidemic that had devastated their lives and killed thousands, perhaps millions, of others. Molly Crosby has provided a brilliant and deeply moving account of the fearful years between 1915 and 1927, when this mysterious, worldwide pandemic struck, giving us vivid, intensely human portraits of seven individuals caught up in this epidemic, and the physicians who did their best to understand and help them. In the end, *Asleep* reminds us that this strange, often terrible disease is not extinct, only quiescent. It may well strike again in our lifetimes." —Oliver Sacks, author of *Awakenings*

"The engaging story of the outbreak of a bizarre disease . . . [Crosby] provides fully realized portraits of not only her case studies' patients, but also the brilliant doctors who treated them." —*Kirkus Reviews*

"Here's medical curiosity combining history, mystery, and riveting storytelling." —*Publishers Weekly*

"Harrowing." —*Discover*

"A moving nonfiction account of a complicated disease."
—*The Dallas Morning News*

"Crosby chronicles the outbreak through a series of horrific and heartbreaking case studies." —*Fort Worth Star-Telegram*

"Crosby follows *American Plague* . . . with the equally terrifying and still unresolved story behind encephalitis lethargica."
—*The Memphis Commercial Appeal*

"There is something extravagant about a good book; it is an indulgent treat. . . . *Asleep* was, indeed, just such a pleasure." —Blogcritics.org

ACCLAIM FOR

The American Plague:
The Untold Story of Yellow Fever,
the Epidemic That Shaped Our History

"Gripping . . . Highly readable." —*Newsweek*

"Engrossing . . . A first-rate medical detective drama."
—*The New York Times Book Review*

"Seamlessly blends history and science to tell us how yellow fever haunted the nation—and why, if we're not extremely vigilant, it will haunt us again." —Hampton Sides, author of *Blood and Thunder*

"A forceful narrative of a disease's ravages and the quest to find its cause and cure. Crosby is particularly good at evoking the horrific conditions in Memphis, 'a city of corpses' and . . . also relates arresting tales of heroism." —*Publishers Weekly*

"A fascinating book about yellow fever, its unspeakable horrors, and the uncommon valor that four doctors displayed in their quest to solve a devastating medical mystery." —*The Tennessean*

ASLEEP

The Forgotten Epidemic
That Remains One of
Medicine's Greatest Mysteries

MOLLY CALDWELL CROSBY

BERKLEY BOOKS, NEW YORK

THE BERKLEY PUBLISHING GROUP
Published by the Penguin Group
Penguin Group (USA) Inc.
375 Hudson Street, New York, New York 10014, USA
Penguin Group (Canada), 90 Eglinton Avenue East, Suite 700, Toronto, Ontario M4P 2Y3, Canada
(a division of Pearson Penguin Canada Inc.)
Penguin Books Ltd., 80 Strand, London WC2R 0RL, England
Penguin Group Ireland, 25 St. Stephen's Green, Dublin 2, Ireland (a division of Penguin Books Ltd.)
Penguin Group (Australia), 250 Camberwell Road, Camberwell, Victoria 3124, Australia
(a division of Pearson Australia Group Pty. Ltd.)
Penguin Books India Pvt. Ltd., 11 Community Centre, Panchsheel Park, New Delhi—110 017, India
Penguin Group (NZ), 67 Apollo Drive, Rosedale, North Shore, 0632, New Zealand
(a division of Pearson New Zealand Ltd.)
Penguin Books (South Africa) (Pty.) Ltd., 24 Sturdee Avenue, Rosebank, Johannesburg 2196,
South Africa

Penguin Books Ltd., Registered Offices: 80 Strand, London WC2R 0RL, England

Cover design by Lesley Worrell
Cover art © 2010 by Axel Kock / Shutterstock
Book design by Kristin del Rosario

PRINTING HISTORY
Berkley hardcover edition / March 2010
Berkley trade paperback edition / February 2011

Berkley trade paperback ISBN: 978-0-425-23873-8

The Library of Congress has catalogued the Berkley hardcover edition as follows:

Crosby, Molly Caldwell.
Asleep: the forgotten epidemic that remains one of medicine's greatest mysteries / Molly Caldwell Crosby. — 1st ed.
p. cm.
Includes bibliographical references and index.
ISBN 978-0-425-22570-7
1. Epidemic encephalitis—History. I. Title. II. Title: Forgotten epidemic that remains one of medicine's greatest mysteries.
[DNLM: 1. Encephalitis, Arbovirus—epidemiology—Europe—Case Reports. 2. Encephalitis, Arbovirus—epidemiology—United States—Case Reports. 3. Encephalitis, Arbovirus—history—Europe—Case Reports. 4. Encephalitis, Arbovirus—history—United States—Case Reports. 5. Disease Outbreaks—history—Europe—Case Reports. 6. Disease Outbreaks—history—United States—Case Reports. 7. History, 20th Century—Europe—Case Reports. 8. History, 20th Century—United States—Case Reports. WC 542 C949a 2010]
RA644.E52C76 2010
362.196'84980094—dc22

2009034928

PRINTED IN THE UNITED STATES OF AMERICA

10 9 8 7 6 5 4 3 2 1

For Andrew, Morgen and Keller

CONTENTS

AUTHOR'S NOTE

By the very nature of this disease and its range of symptoms, some of the stories presented in this book seem unbelievable, even impossible to imagine. Nonetheless, this is a work of nonfiction.

This disease is known as "the forgotten epidemic" and for good reason. At the time I was researching and writing, not a single contemporary book on the epidemic existed. Part of my research included an interview with neurologist and writer Oliver Sacks, whose powerful book *Awakenings* is the most famous story stemming from the epidemic, though the focus of his account is a group of patients, survivors of this forgotten epidemic, he encountered in the late 1960s. Sacks himself was surprised to find these victims of the disease languishing in mental institutions. He wrote, "I would not have imagined it *possible* for such patients to exist; or, if they existed, to remain undescribed."

The bulk of my research for this book came from sifting through old medical journals and case studies from the 1920s, primarily those in the archives of the New York Academy of Medicine and Columbia University's Augustus C. Long Health Sciences Library. All the case studies in this book are based on actual medical records and articles published in medical journals from the time period, as well as personal accounts, family letters, and newspaper coverage.

In accordance with the Health Insurance Portability and Accountability Act (HIPAA), as well as respect for patient confidentiality, names and personal details that would point to patient identification have been changed. In cases that were covered in major newspapers and documentaries, names and personal details were *not* changed, as their cases were publicly documented and historically significant to this epidemic.

Each of the cases was chosen to represent the range of symptoms from this kaleidoscope of a disease—some slept, some died, some recovered, some were crippled—but it is the collection of their cases that is the story. It was an epidemic that was shocking not because of the number of victims or medical breakthroughs or famous physicians, but because it so horribly and hauntingly altered the lives of its patients and the doctors who tried to save them.

This disease may have existed for hundreds of years, epidemics of it weaving in and out of the seam of history, but one amazing fact seems to hold true: physicians are surprised and confounded by it each time it appears, from the 1920s epidemic to Oliver Sacks's research in the 1960s to the cases discovered today. It begs the question: should this disease return in epidemic form, will we be just as bewildered and unprepared as those who came before us?

Though cases have become rarer within the last few years it is unlikely that it will ever again vanish from medical memory.

—DR. CONSTANTIN VON ECONOMO, 1931

Such forgettings are as dangerous as they are mysterious.

—DR. OLIVER SACKS, 1973

PROLOGUE

Inside

My grandmother was sixteen when she fell asleep.

What she remembered most from those weeks was the whiteness and the emptiness. Not whiteness like snow or fresh paint, but one caused by the complete lack of anything else, the same way a dense fog can consume everything in its path. She also felt cool, like the summer nights when she slept with a bowl of ice in front of the fan. Everything seemed cold and vacant and white.

She could see herself present in the room, but she was not herself. She was polished somehow. Smooth and sculpted like a statue. She tried to lift her arm, but it would not move. She concentrated and tried again, but she felt as though her arms, hands, legs, and feet were no longer connected to her brain, no longer accepting commands. At that point, she became frightened. An overwhelming, claustrophobic fear seized her: she *was* a statue.

As though her mind could only take so much of this stress, she began to pull away, teetering on the edge between a dream and wakefulness. Finally, her mind told her that she was only dreaming, it was only a nightmare, and she felt the relief rush through her like a winter breath. She began to feel her bedsheets on her skin and sensed the sallow light from her eastward window. She struggled to come out of the dream and open her eyes, but it would not happen. The feeling that something terrible was happening, a taste of the nightmare, remained.

There were things she knew even through her closed eyes: in the morning, the drapes opened and the lamps were extinguished, the smell of gas still streaming from the wick. When the drapes closed again in the afternoon heat, it was like a heavy cloud passing in front of the sun. She could smell the purple, bearded iris that her mother kept in a vase on the piano. And she remembered how she ended up in bed, when drowsy, feverish, and weak, she had fallen down the stairs of her parents' home.

There were things she could not know with her eyes closed: the passing of time. Time, after all, is a fixed point in life we take for granted. To lose it is like losing balance. She could not feel herself, or at least how to control herself. Her body no longer answered to her mind. She was imprisoned by bars of bone in a windowless cell. And she did not know the strange voices that came and went from her room, mingling with the sound of her parents' voices. She heard them discuss how many weeks of school she had missed; she heard them say "*sleeping sickness*."

The year was 1929, and at the time, Virginia and her family had no way of knowing she was joining millions of others suffering from a strange, global pandemic—a disease that would change medicine itself, but vanish from medical history. A disease that would kill close to a million people and leave thousands more languishing in mental institutions for the rest of their lives. An epidemic that nearly a century later remains a mystery, but could strike again.

* * *

Virginia's mother spoke to her in that calm tone of voice used for bloodied shins or turned ankles, and her room was filled with the muffled voices of doctors. What was most frightening was the uncertainty in their voices. They did not know what had caused this in a healthy teenage girl, and worse, they did not know how to stop this endless sleep. Her temperature was taken several times and noted on her chart. She felt the doctor's hand against her wrist and on the stem of artery along her neck. And then, over the course of the days and weeks, she heard the doctors pronounce her dead—three different times. Each time, she listened to her parents weep and heard them make plans for her viewing and burial.

She could not even tell them they were wrong.

CASE HISTORY ONE

Austria, 1916–17

NAME: An unknown soldier

PHYSICIAN: Dr. Constantin von Economo

CHAPTER 1

An Epidemic Begins

From miles away, the Earth shook. Starbursts of white light exploded overhead. Rockets whistled, and exploding shells rumbled like thunder. An enormous mass of flint-colored smoke cloaked everything. As many as 100,000 shells fell like rain every hour, leaving pieces of metal all over the fields to shine like silverfish in the moonlight. Where the shells landed, geysers of mud flew, stones and splinters sprayed, and the trenches, in long lines, like scars in the landscape, would break, leaving men abandoned all over the field. Lying in the hollowed-out trenches, separated from the rest of the lines, looking to the sky, the French soldiers watched as German observation balloons flew overhead. A French carrier pigeon, their last one, flew through the cloud of smoke carrying dispatches: "We are still holding our position, but are being attacked by gases and smoke of very deadly character. We are in need of immediate relief." A few days later, another message from a dispatcher read, "We are in desperate straits." And, finally, a fragmented note that said only: " . . . must go on."

The battle of Verdun, fought in 1916, is regarded by many historians as the longest and most devastating battle in world history. The plan was born on Christmas night, 1915, when German command decided to attack the ancient French fort town in a battle that would come to epitomize a war of attrition. It was meant only to bleed the French army white. No ground would be gained or lost or even sought. The battle would simply match life for life until the army with the fewest souls left surrendered. The pointless slaughter lasted ten months with over 700,000 casualties.

For men who had been raised believing war to be a venture in heroism, patriotism, and bravery, the sad truth became all too real during World War I. Young men who dug deep into muddy trenches, surrounded by barbed wire, were of little use against the machinery of an approaching enemy. What little hand-to-hand combat the men fought took place amid a spray of machine gun fire in a vast openness between the trenches called "no-man's land." To most soldiers, there seemed to be neither heroism nor cowardice in hunkering down in a trench waiting for a shell to fall and explode. Many wounded were left on the field to die because many more would have died retrieving them. Bodies that fell remained until the rats found them, and rather than exterminate the rats, the armies decided to let them do the cruel work of returning to the Earth what belonged to her.

Any photographs from World War I are, of course, in black-and-white. They appear in shades of gray more than anything else. The bodies are hard to discern from the mounds of dirt, decapitated tree trunks, and upended hedges, bare and spindly like beached coral. The faces of the living and those of the dead are also hard to separate. In every direction soldiers looked, green landscapes had turned sepia-toned, golden fields had grown ashen. For the soldiers, too, the world became colorless—even the blood, which ran so heavily it turned black.

It was a war of unnerving, even surreal contrasts, with bayonets and machine guns, carrier pigeons and aircraft, cavalry charges and tank warfare. Horses and men alike wore gas masks in chemical battles. This was not a war the world was prepared to understand, a conflict born out of the Progressive Era when technology changed everything, medicine had made tremendous advances, and humanity had become in every sense of the word more civilized. When that technology and science were used against people in the form of weaponry and chemical warfare, the human psyche was assaulted as well. That realization would come later—in the offices of neurologists and psychiatry clinics.

The men who entered the war, the soldiers who were capable during Christmastime of 1914 to come out of the trenches to meet the enemy on the field, sing "Silent Night," and exchange cigarettes, would no longer exist by 1918. In those four years, violence, terrorism, genocide, attrition, and civilian casualties were seeded and grown in the twentieth century. World War I took every hope for humanity and shattered it. But the battle of Verdun would leave the world with another legacy not yet realized.

Army ambulances, some still horse-drawn, flanked the edges of the battle, with medics and stretchers waiting for a clearing. Screaming voices could be heard even above the artillery and lisping sound of stray bullets. When the smoke cleared enough, the wounded soldiers who could be reached were carried from the chaos of the battle to make the short and rutted ride in an ambulance to the field hospital. There, in neat rows, contrary to everything on the battlefield, modes of injury and death were laid out in an organized grid with as much exactitude as the tombstones in a field.

At least one wounded patient made this journey from the trenches to the field hospital, where he waited for the medical train. In the beds all around him were injured or dying men. Some were dying of wounds inflicted by men—blinded by mustard and chlorine gas, whole faces missing, limbs lost to the constant shelling. Some bodies were broken open, peppered with shrapnel. Although field hospitals had no shortage of gruesome injuries inflicted by machines, the ones inflicted by microbes proved even more deadly. The victims of those enemies had none of the crude, gaping wounds of field injuries. The enemy that attacked these men killed from within. The epidemic diseases had their own wards, with beds separated by white curtains to help contain the spread of illness. After all, battlefields and war hospitals were aflame with typhus, cholera, influenza, pneumonia—especially as war rations had produced emaciated, malnourished soldiers. It's a strange fact of war that a conflict between men becomes more than a contest with weaponry, but a conflict within nature as well. In the most practical sense, city men clash with country men, western men with eastern men, northern men with southern men. Thousands of people who would not normally encounter one another are brought from all corners of the world. Some were brought from the Far East to help build trenches. Some were brought from across the Atlantic to fight. Others were shipped from colonies in Africa or Southeast Asia. Unknowingly, they brought some native microbes as well. The war presented nature with its own unique opportunity, a microscopic holocaust.

This unknown soldier may have had a physical wound or an illness or even a mind that had been injured. Whatever the reason, he was evacuated from the hellish battlefield at Verdun for the field hospital and, finally, for Paris.

Two physicians on opposite sides of the war worked in neuropsychiatric hospitals—one clinic was in Paris, the other in Vienna. The psychiatric clinics filling with patients were testament to the brutality of this war and the inexperience of its participants.

When the war began, soldiers still wore leather helmets, but they soon learned that modern warfare left its mark by the metal embedded in skulls, as well as the cuts inflicted on the psyche. By the end of the war, helmets were made of steel. Even with the stronger helmets, shrapnel could cause visible injuries to the brain and, inadvertently, gave neurologists a chance to map the brain, matching its parts to body functions, using physical damage as the guide marker. Likewise, the psychological damage was paving the way for modern psychiatry. In what would become one of the few positive points to come from World War I, brain study would advance at rates faster than ever before.

The French physician, a pathologist and pediatrician, working in the Paris clinic was named Jean René Cruchet. When Cruchet met with the unknown soldier from the front in Verdun, he was struck by an unusual set of symptoms. At first, he wondered if the symptoms could be the aftereffects of mustard gas or even a new chemical weapon. Other patients soon followed, and in all, Cruchet saw sixty-four similar cases with no standard diagnosis. What was striking about these patients was the unusual range of symptoms. Some had fever; others did not. Most complained of headache and nausea. Strangest of all was how much these soldiers slept.

It would have seemed almost serene at first—blank, expressionless faces, free of terror and pain, calmly asleep in row after row after row. What was frightening about these men was that they would not wake. It must have felt like being in a room full of the breathing dead. At first, their symptoms vaguely matched any number of diseases, but as they progressed, they did not follow the course of other diseases of the trenches. The soldiers were not comatose; they were simply asleep.

Cruchet spent several months studying the sick soldiers, and he prepared a paper on the unusual cases he found coming from the trenches in Verdun. He did not know, could not know, that on the other side of the war another physician was witnessing the same. That physician, in a clinic in Vienna, was preparing a similar paper on the same subject. Though the two men could not coordinate their efforts, nor even contact each other, they were witnessing the same disease. By definition, that made it an epidemic.

CHAPTER 2

Constantin von Economo

A drowsy, confused patient wandered into a psychiatric clinic in Vienna in January 1917. He sat slumped over in his seat, nodding off to sleep, his neck falling forward or his head hanging limply to the side like the wing of a wounded bird. He could be roused, but not fully awakened, peering out from a sliver of open eye. Aside from his obvious sleepiness, the most unusual thing about this patient was that he was a civilian.

The Wagner-Jauregg Clinic had seen a large number of patients recently, but most of them were returning soldiers with wounds to the head. This sleepy, lethargic patient had nothing to do with the war. The physicians could not even identify what was broken—there was no physical trauma, no history of insanity, no single infection that caused symptoms like this. It was unlike anything the doctors had ever seen. To make matters worse, his case was soon followed by a dozen more.

In nearly all cases, the patient's eyes changed. The eye muscles malfunctioned, causing dull, unresponsive eyes, or just the reverse: eyes that twitched or rolled back into the head. So many unusual symptoms appeared in these cases that in a timely analogy one doctor compared it to a burst of shrapnel in the nervous system.

The symptoms only became stranger.

An epidemic of hiccups erupted, and one of the clinic's patients actually died of incessant hiccuping. Other patients exhibited bizarre tics—repeating the same word or phrase, blinking uncontrollably, twitching. There was also a sudden influx of schizophrenia patients, dementia patients, patients who salivated uncontrollably, or ones who froze still in a catatonic state. In all cases, an unusual sleepiness accompanied the symptoms.

It was a complete mystery to the physicians at the clinic. The only factor connecting these patients, the one thing they had in common, was no clear diagnosis. In medicine, and in brain study in particular, that created an unusual dilemma—this was a disease that could be diagnosed only by ruling out all other diseases.

In this clinic in Vienna a young, unknown neurologist named Constantin von Economo first took notice of these peculiar patients who seemed to have no connection to one another and found the link common to all. There was no known neurological disease that caused indefinite sleep. And, likewise, with no other disease would the patient suddenly and inexplicably awaken.

"We are dealing with a sleeping sickness," he said, bewildered.

Dr. von Economo waited in the halls of the Vienna psychiatry clinic. They were white and stone. Clean. They reflected lamplight and smelled of antiseptic. Snow fell outside the windows. Indoors, however, the clinic was teeming with physicians and patients. Long metal tables were stacked with paperwork, charts, and graphs. It was orderly and efficient.

Von Economo could not have been better positioned. In Vienna, and throughout eastern Europe, groundbreaking work on brain study was taking place during the early decades of the twentieth century. Julius Wagner-Jauregg, the neurologist for whom the Vienna clinic was named, was experimenting with fevers to treat mental illness. He would win the Nobel Prize for his experiments using malaria as a treatment for neurosyphilis. In neighboring Germany, Emil Kraepelin was studying dementia praecox, the disease that would later be called schizophrenia, and Alois Alzheimer was publishing papers on the disease that would bear his name. And Europe was becoming the birthplace of psychoanalysis, thanks to Sigmund Freud and Carl Jung. According to Freud, exposure to so many taboo, immoral, violent actions was more than the human mind could take in many instances. The scores of patients with "war neurosis" seemed to support his theory as patients became catatonic, suffered nightmares, and saw recurring images of bayonets and mutilated bodies. They could not get the scent of the yellow clouds of mustard gas out of their nose, the feel of trench barbs out of their skin, or the staccato sound of continuous gunfire out of their ears.

Von Economo himself was more of a clinical physician and had returned to Vienna only a few months before. Like most other Austrians, he had enlisted in the weeks after Austria's archduke had been assassinated by a disgruntled patriot from Sarajevo. He enlisted first for the air corps, but when he was stationed in Vienna to organize the air force, he transferred to the automobile corps to get onto the Russian front faster. While the Central Powers had declared the war would be over by Christmas, von Economo rightly viewed that as a gross underestimation. "The war is a huge wave," he told his wife, "that will not ebb for a long, long time." Eventually, waiting out the long winters in small villages along the Eastern Front bored von Economo. He reapplied for a position in the air force in 1916 and was at last granted one. On the days when he was not flying, von Economo volunteered to do medical assistance.

In fact, medicine was a second career for von Economo, and he continually turned down promotions to pursue other interests. His first love was flying—in balloons, then in planes. He was the first in Vienna to attain an international aviation license. Von Economo also loved anything to do with science and learning in general. He had received a degree in engineering before achieving one in medicine. He was something of a left-brained Renaissance man—the kind of scientist who kept copies of the New Testament, the *Odyssey*, and Faust on his bedside table.

As for his professional life, von Economo didn't really need one at all—although born in Romania, von Economo came from Greek aristocracy and married the daughter of an Austrian prince, becoming Baron Constantin von Economo. He lived a cosmopolitan life, traveling between estates in Italy and Greece and his home in Austria. His appearance was a nod to his Greek heritage—chestnut skin, dark hair parted and swept to one side, black eyes beneath heavy lids, a thick mustache that turned at the edges. He looked calm and dignified, a man who would be at ease in any company—whether European royalty or patients of a psychiatric clinic.

Von Economo spent months flying with the Austrian air corps, but when he learned of his brother's death in battle, he returned to Vienna to be with his grief-stricken parents. At their behest, he reluctantly transferred to a medical command in the city away from danger. When his wife asked him if he missed aviation, he thought for a moment and replied, "No, there is no longer so much new to do there." The psychiatry clinic, and medicine as a whole, still provided plenty of uncharted territory. It was at the Wagner-Jauregg clinic that von Economo met his first sleeping sickness patient.

On that cold winter day in Vienna, in a hospital that looked out upon the graying snowcaps on rooftops and a sky laced with chimney smoke, the clinic was filling up with these patients literally falling asleep where they stood. Family members complained that a patient might fall asleep at dinner with food still in his mouth; or parents would shake and shout at a child who would not wake from her sleep.

Before long, the novelty of this disease evolved into something much more frightening—some of the patients fell asleep indefinitely, eventually dying. The mortality rate of this strange new disease was as high as 40 percent in the cases at the Vienna clinic. When the doctors autopsied the first victims, they found significant swelling and damage to the midsection of the brain—the part of the brain that controls sleep.

With a clinic full of patients dying from sleep, von Economo searched medical texts for clues to what this epidemic could be. The culprit in any disease outbreak usually falls into one of three categories: an entirely new pathogen, a pathogen that has been around for some time but was only recently identified, or an old pathogen that has adapted and grown more dangerous.

Von Economo could find nothing like this sleeping epidemic in any of the standard literature. Some papers mentioned a mass suicide in Scotland that pointed to a sleeping sickness disorder followed by hysteria. The patients, as a group, jumped from their windows. Similar instances of collective hysteria surfaced in history, all related to sleep patterns. And von Economo did have a vague memory of an epidemic of sleep, called "*nona,*" when he was a child. No medical texts describing the outbreak were widely published, but it was recent enough still to circulate in the memory of the previous generation. The sleeping sickness had occurred in Italy and other parts of Europe during the 1890s just after a notable flu epidemic. Von Economo asked his mother, who remembered when the nona epidemic struck in western Europe, and she told him in detail about the symptoms—they sounded hauntingly similar to what he was seeing in his clinic in Vienna. The peasants called the disease "La Nonna," which means "the grandmother" in Italian, but in the cities, the disease became known as nona, or "the living dead." At the time, it was thought to be a neurological complication of the flu.

The nona outbreak became the first time doctors questioned whether or not influenza and this sleeping sickness were somehow connected, whether the flu had led to these unusually lethargic patients, or if this strange new disease was an epidemiological foreshadowing of an influenza pandemic. More than one hundred years later, the question would remain.

Von Economo quickly pulled together a paper to present in spring to the Vienna Psychiatry Society, and on April 17, 1917, he announced that he had identified this new disease. His paper was published within days of the one published by the Parisian surgeon, Jean René Cruchet. In Vienna, the disease would become known as von Economo's encephalitis; in France, the disease was named for Cruchet. To the world at large, a less controversial name developed that meant, literally, a swelling of the brain that makes one sleepy, or *encephalitis lethargica*.

Of course, there was some opposition in Vienna to von Economo's theory of the disease, particularly among psychoanalysts, who believed mental illness to be a result of deep-seated psychological problems, not an actual disease. But von Economo warned that any psychiatrist who based his studies entirely on the subconscious, ignoring organic causes like disease, would build his theories on sand and watch the "psychological constructions collapse like a house of cards."

Von Economo began to discern a pattern of brain damage in his autopsies, and he suspected a virus was to blame. Before the age of the electron microscope, there was no way to identify a virus aside from infecting another animal and awaiting the outcome. He found that brain tissue of the encephalitis patients could pass the disease into monkeys, producing similar symptoms and suggesting that the culprit was both microscopic and virulent.

Von Economo also made an important discovery about the brain. He theorized that a certain part of the brain, the hypothalamus, was regulating the sleep in his encephalitis patients. The hypothalamus, small, almond-sized, and located in the middle of the brain, controls several automatic functions like body temperature, hunger, thirst, and sleep cycles. Depending on where the encephalitis lethargica damaged the hypothalamus, too much sleep or too little resulted. He believed the hypothalamus was the part of the brain responsible for two other known sleep disorders as well: narcolepsy and insomnia. Von Economo's theory was visionary to say the least; seventy years would pass before advanced medicine could prove his theories true.

As the European doctors squabbled over what was causing encephalitis lethargica, how it spread, and who discovered the first case, the disease trumped the scientists. It vanished. No new cases appeared. The doctors felt relieved, fortunate even, that this seemed to be the end of a terrifying epidemic. This epidemic had been so frightening not because of the high death toll or the clip at which it spread, but because physicians had so few answers. At the same time, they must have felt some disappointment. The disease had come and gone before they ever had the chance to solve the mystery—it was a feeling that would be echoed again and again.

In Europe and elsewhere, few people worried about the little-known epidemic of sleep. They were preoccupied with the rapidly spreading, extremely virulent influenza breaking out among troops in Europe and the United States. There is no definitive calculation, but the number of people about to die of influenza in 1918 would fall between 20 million and 100 million worldwide. The sleeping sickness faded quietly behind the voracious flu pandemic, a drop of water in a tempest. Surely, encephalitis lethargica was a bizarre and isolated event, like the previous ones in history. A sudden, mass hysteria. Short-lived and soon forgotten.

Then, an ill patient—delusional, feverish, and lethargic—arrived at London Hospital.

CHAPTER 3

The London Outbreak

World War I, though sparked by an assassination, was really a war fought over alliances and power struggles. Two countries, Austria and Serbia, went to war. Thanks to complicated, and sometimes outdated, alliances, nine other countries went to war as well. Britain had an old but steadfast agreement to ally with France in the case of a war. So, during the first two years of World War I, Britain sent a respectable number of troops into Belgium and France. It wasn't long before the Western Front became overwhelmed, and the British engaged fully.

As disastrous losses at Verdun took their toll on the French soldiers, Britain decided to send her own troops into northern France to relieve some of the pressure by distracting Germany in battle. The British and German troops met on a field near the river Somme in a battle that cost Britain more losses in a single engagement than in any other war. The British plan was costly, but ultimately effective. British troops finally began to overwhelm Germany.

In a fateful turn of events, legend has it that a British private came upon a wounded German courier during the battle. In the face of defeat, the courier did not even reach for his rifle. Looking him directly in the eye, the British soldier could not bear to pull the trigger on a wounded man in the field. Sympathy and respect for life overtook him, and he lowered his rifle. It would prove to be a regrettable decision; the German courier's name was Adolf Hitler.

Hitler was hospitalized both after the battle near the Somme as well as during a later engagement. During a hospital stay toward the end of the war, his erratic and volatile temperament caught the notice of physicians and patients. Most of Hitler's medical records were later destroyed, so no complete list of his symptoms or diagnosis exists, but several different diseases have been suspected: the whole spectrum of mental illnesses, syphilis, hysterical blindness, mustard gas poisoning, and encephalitis lethargica. Though it seems unlikely that any medical diagnosis can ever explain evil.

As Britain's involvement in the war escalated, and more and more troops landed along the coast of France, injured or ill British soldiers on the front were sent home—by train to the coast then aboard the hospital ship and back across the English Channel. With so many troops traveling back and forth to England, every effort was made to prevent soldiers from carrying disease with them as well. Sick soldiers in the field hospitals had their uniforms removed and disinfected before they boarded the train for the coast. Once across the English Channel, the soldiers carrying infectious disease were sent to the clearing hospital to wait out the illness before they returned on sick leave to their homes in London or elsewhere in England.

Nonetheless, the exchange of troops became an exchange of microbes as well. As early as 1916, a British doctor named A. J. Hall read reports of a new illness spreading among the troops in France—

one that sounded very much like what von Economo and Cruchet had reported.

The path between Verdun and the Somme was but one of many troop movements and battles that gave soldiers the chance to spread this sleeping sickness. The troops crowded together in camps, then trains and boats, giving the disease a chance to broaden its reach and offering the perfect entry in the United Kingdom. One ship after another departed from the coast and crossed the ragged whitecaps between France and England.

When patients with drooping eyelids, slurred speech, and muscle weakness first began filling the waiting rooms of London clinics, sausage was blamed. With a wartime shortage of meat, questionable substitutes were often used, so botulism was the first diagnosis. Other foodborne toxins were investigated as well. To complicate things further, families who *had* shared the same food did not see multiple cases, but infants, who were exclusively breastfed, with no exposure to tainted foods, *were* experiencing the strange symptoms. When London Hospital failed to find the bacterium that caused botulism, it put an end to the botulism scare. After that, tear gas and mustard gas were suspected, or even a new type of chemical warfare. Finally, British physicians simply declared this to be an entirely new disease.

It would have been helpful to the British doctors to have some contact with Austria or even French physicians at this time and to know about the epidemic of sleep that had arisen and evaporated, but the war severed most medical communications. The war was disseminating the disease, but not the information about it. Von Economo's paper, as well as the one published by Cruchet, received little attention across the Channel. Physicians were too overwhelmed with war injuries and fatalities to pay much attention to anything else. Yet encephalitis lethargica, which had taken root in Europe, was spreading as steadily, slowly, and unnoticed as a creeping vine.

Working with the Ministry of Health, the British physicians pulled old medical files, books, and published papers to search for information; they reviewed historical documents. Eventually, they traced cases of unexplained sleep as far back as 1657 in Copenhagen, 1658 and 1661 in England, and 1775 in London. In fact, those three epidemics of sleeping sickness during the seventeenth century may very well have been the inspiration for one of our most famous fairy tales: "Sleeping Beauty," published in 1697. Likewise, Washington Irving wrote "Rip Van Winkle" while he was living in England in 1819, and it, too, is the story of someone awakened after years of sleep. The mid-1800s in Germany, physicians reported, seemed particularly rife with paralysis or coma caused by fevers. It was around that time that Edgar Allan Poe wrote two of his most haunting tales. In both stories, a character falls in and out of a catatonic, "death-like" trance from which he cannot be awakened, and Poe plays on one of society's greatest fears at the time. One short story was "The Fall of the House of Usher"; the other was suitably titled "The Premature Burial."

The first cases in Britain proved as idiosyncratic as those in Vienna. The English general practitioner A. J. Hall compared it to the story of Cinderella: "They could not find amongst the diseases of their acquaintance a foot which would go into this newly found glass slipper." Hall's first case, a boy, did not seem particularly sleepy; but he did have marked inability to move his muscles. He could barely lift an arm or leg off the bed. Other patients of Hall's had fevers for weeks at a time or mumbled incoherently or had tremors. Most of the patients, however, showed the same lethargy. It was as if their sleep cycle was turned around—they stayed in a sleepy stupor all day and became delirious at night. Still other patients could sleep with their eyes open, with the chilling appearance of a corpse. Worse, the cases in England were showing a mortality rate over 50 percent, even higher than the death rate in Vienna. Death, the Ministry of Health concluded, appeared due to paralysis of the respiratory system. In 1918, the first year the disease appeared on English soil, there were 538 cases; the number doubled in the coming months. The Ministry of Health also made it clear that "Medical Health officers are invited to take any action which they consider useful and practicable to secure post-mortem and pathological examinations by skilled workers who are investigating this disease." Public health was of greater importance than individual wishes during an epidemic. The ministry made encephalitis lethargica a "reportable disease," forcing physicians to inform the government of new cases.

As the disease spread, however, it took on new symptoms.

In a bizarre twist on the lethargy cases, some patients grew hyperkinetic, unable to sleep. Their bodies twitched and muscles tensed. A broader range of uncontrollable tics occurred. If put to bed, a patient might begin rolling from side to side. A heightened sense of euphoria was reported in some cases and incessant talking in others. In one report, a boy was described as leaping from all fours into the air—like the mechanical toys sold on the London streets called "jumping beetles." It would have been humorous had it not been for the look of terror on the boy's face. He could not control himself or stop. As the number of cases increased, one British physician noted that with encephalitis lethargica, recovery, in the true meaning of the word, was becoming the exception rather than the rule.

For the first time physicians not only identified the disease outside of Vienna or Paris, but they also realized the symptoms were changing as it spread. By 1918, doctors in England noticed with alarm that the epidemic was exceeding the severity of the previous years. This would prove to be a foreboding discovery.

What began hesitantly as an epidemic in the trenches of French war fields was now becoming pandemic, reaching countries all over Europe. As the United States sent troops over to Europe, the path for the disease broadened once again and struck the one city from which all U.S. soldiers departed and returned: New York.

Encephalitis lethargica would soon spread to countries covering the globe, from Sweden to India, Egypt to China, Australia to Algeria, Uruguay to Persia. And, yet, there were still far more questions than answers about this global disease. In what proved to be unprophetic language, A. J. Hall wrote: "It may be that generations which follow us will see clearly where we can only grope darkly."

Physicians like Hall, in all their frustration, must have had some faith in future medical discoveries and advancing technology. At one hospital, Queen Mary's in London, pathologists autopsied some of the fatal encephalitis lethargica patients. They took tiny pieces of brain tissue, soaked them in formalin, and then coated them in wax. The samples, which looked like clouded pieces of ice with a kernel of brain matter in the center, were then placed in a wooden crate, which was stamped and marked as a postmortem sample from that year: *P.M. 1918*. Each crate was carried to the basement and stacked on shelves holding other boxes marked the same way, dating all the way back to the turn of the century. There it began, slowly, to collect dust for the next eighty years.

Epidemiology literally means "the study of what is upon the people," and in particular, it implies investigating an outbreak of disease. The complex study includes a cross section of several factors that may or may not seem related to disease, like environment, genetics, food and water sources, seasons, contact with animals and insects, or even the way people bury their dead. It requires a mind that can collect minute details and data, while also being able to see the larger patterns they create. An epidemiologist is interested in not only the behavior of a disease, but also the behavior of its victims. Above all else, what will set a talented medical investigator apart from others is the ability to challenge deeply entrenched theories and see the things others have missed. Often, people are blinded by their established beliefs; so much so that they become oblivious to the obvious.

Hippocrates is usually considered the first epidemiologist. He made the connection between illness and human behavior, finding a kinship between the diversity of life and the diversity of illness. An epidemic, he observed, is just the opposite: "When a large number of people all catch the same disease at the same time, the cause must be ascribed to something common to all. . . ."

Notable examples of epidemiology include John Snow's mapping of the cholera outbreaks in London around a contaminated water source; Robert Koch's demonstration that a germ was *causing* diseases like cholera; Walter Reed's work with typhoid in the army camps and his demonstration that mosquitoes spread yellow fever; Sara Josephine Baker tracking the case of Typhoid Mary; and Constantin von Economo's discovery of sleeping sickness.

Identifying the pathogen or its spread is not what led these epidemiologists to triumph—it was their ability to understand the disease's relationship to a place and its people, because epidemiology's long shadow inevitably falls on public health. After all, the work of an epidemiologist often has the greatest impact on cities and communities, and the people in those neighborhoods will live or die by the epidemiologist's success or failure. A true medical investigator has to understand both the organism beneath the microscope and the larger organisms that make up a population. And so the story of an epidemic disease is really just the breakdown of organisms: those that cause the disease, those that suffer the disease, and those that adapt to it.

* * *

In 1919, in New York City, a large number of people all caught the same disease at the same time. The medical investigators did not know the common cause, and they could not see the microorganism beneath the microscope. But they witnessed firsthand, in frightening numbers, those suffering the disease.

The epidemiologists began looking at the larger organisms in the equation: the victims. They looked for connections among people, in neighborhoods, in what they ate or drank, in how they lived, in their behavior. They examined the relationship between the disease and its victims and the kinship between those victims and the place in which they lived.

If an organism is truly a living system that grows and changes in order to maintain a stable whole, there was one more large, complex organism in the story of this epidemic. It was the city of New York itself.

CASE HISTORY TWO

New York City, 1918

NAME: Ruth

PHYSICIAN: Dr. Frederick Tilney

CHAPTER 4

New York City

New York grew faster than any of the world's largest cities—cities that had centuries and all of history to build their empire. It would grow from a sliver of island where ship masts towered over squat buildings to one with a mountain range of skyscrapers. The science of life and the science of a city are not very different, and a city must adapt in order to survive. New York City did more than just survive; it thrived. The city flourished from its earliest exploration to the twenty-first-century giant it is today.

Iconic New Yorker E. B. White wrote that New York should have "destroyed itself long ago, from panic or fire or rioting. It should have perished of hunger when food lines failed for a few days. It should have been wiped out by a plague starting in its slums. . . . " But it did not. The city adapted to and survived each trial.

The city, wrote White, is "both changeless and changing."

If any one characteristic defined the city of New York, it was the ability to defy boundaries—below ground, above ground, under the water, and into the sky. A modern-day historian wrote that New York was "perhaps the one place in the world where the hand of man shaped the environment as much as the hand of God."

At no time was New York's progress more obvious than during the first decades of the twentieth century. The five boroughs known today as Brooklyn, the Bronx, Manhattan, Queens, and Staten Island consolidated in 1898 to become one great municipality: New York City. That fusion implied something remarkable—that New York was redefining itself, not only in terms of politics or demographics, but also in its composition.

In the years immediately following consolidation, at the turn of the twentieth century, the world changed rapidly as the Victorian Age gave way to the Progressive Era, and New York would undergo a metamorphosis not only in the way it evolved, but in the way the city looked and felt. New York became more than just a place; it became an inspiration. The island city inspired countless artists, writers, architects, and musicians, who flocked there. The Jazz Age was born in New York. In a sense, jazz was an homage to Harlem and the city itself, an eclectic, complex interweaving of cultures. A pulsing sound that could be both chaotic and brilliantly synchronized at the same time. And it was not just musicians who found the rhythms to capture the city, but also artists, photographers, architects, and writers. Their rapture over the city was obvious in the words they chose to describe it: colossal, magnetic, astonishing, feverish, glittering, an imperial city, the Niagara of American life. New York was even described as poetry itself.

But New York also faced a dilemma most American cities did not—identity. As one historian pointed out, New York City was looking east to Europe more often than west to the rest of the nation. Broadening that divide even more was the fact that 40 percent of the population was from various immigrant, often European, backgrounds. And, as a city gifted in many ways, it was difficult to characterize New York too narrowly—whether by architecture or theater or publishing or finance. As a testament to the multifaceted surface of Manhattan, one guide to the city broke down the "City of Cities" into various sections. At the tip of the island was the "City of Banks," "City of Fish," "City of Coffee Dealers," and "City of Gems," among others. Moving north, the jigsaw of districts included the "City of Publishers," "City of Artists," "City of Theaters," and "City of Builders." At the farthest end stood the "City of Hospitals."

What's more, the Progressive Era itself was breaking boundaries in a city that had always pushed forward more quickly than any other American metropolis, and technology was altering the way people lived, faster than it would in any period past or present. The cities that changed with the times would thrive; those that did not would begin to die. It was an urban survival of the fittest.

And so by 1918, New York City found itself at a tipping point. The world had been caught in the grasp of the Great War—the momentum of progress gathered speed in New York City just as major European cities turned their focus to war. By the end of that war, New York had replaced London as the financial center of the world. It had replaced Boston as the publishing capital of the United States. With ties to Paris strained during the war, New York department stores and Fifth Avenue shops became the best shopping in the world. And Harlem and the nightclub scene began to place New York at the focal point of modern music. In the coming decade, the introduction of radio would lend a voice to the vibrant image of the city. Radio, broadcast almost entirely out of New York, would carry the stories of the city's speakeasies, gangsters, socialites, millionaires, politicians, artists, and musicians to the rest of the country.

As the great decade of the 1920s approached, that evolving, modern metropolis was becoming the golden city F. Scott Fitzgerald would write about. New York has often been described as a city of light and dark, and during that time period it was radiant—the same way trees appear brighter when backlit by a storm.

A wind, cold as quicksilver, swept the length of the sidewalks in New York City. Pedestrians hovered against the walls of buildings and held tight to their bowler hats. The 1917–18 winter was the coldest New York had seen in a generation. In December, all records were broken when temperatures dropped thirteen below zero. Snow blanketed the city by January, killing a large number of trees in Central Park, their ice-laden limbs snapping and cracking. The cold fractured railway lines, froze the switches, and turned coal to solid blocks of black ice. It wasn't just the trains that were stalled, but also the ships. Dozens sat in the harbor like metallic icebergs bobbing among the floes.

As the temperatures plummeted, coal rationing started. Coal had already been a scarce commodity during the war, but with few shipments arriving in New York, and with those that did encased in ice, the situation was becoming dire. "Workless Mondays" were issued to keep most businesses, including the New York Stock Exchange, closed one additional day per week to prevent further use of coal.

Dr. Frederick Tilney walked out of his office one icy day in the middle of January. A colorless sky was punctuated by milky clouds and white gulls hanging on the wind over the Hudson River. Tilney always visited patients in the morning so he could keep his afternoons and evenings free for research.

He left his townhouse on Fifth Avenue, stepping through piles of gray snow and splintering frozen mud puddles every few steps. On the sidewalks, snow had been collecting and graying, and icicles hung from shop awnings. The memory of a fresh snow was now replaced with the reality of piles of slush fouled by ash, horse manure, and litter.

In spite of the strong scent of refuse, the harsh cold only sharpened the smell of roasting chestnuts in the handcarts, coffee at the lunch counters, tobacco smoke, and coal-burning fires. In the distance, lonely ship sirens sounded, and church bells marked the hour. The snow had quieted everything, and the city itself seemed to absorb all sound and motion.

Tilney walked past five-and-ten stores, United tobacco, hand laundries, shoe repairs, millineries, drug and soda shops, linen and handkerchief stores, tailors. But the cold air had closed most of the newsstands and the carts selling wares.

As he stood at the edge of the street, Tilney paused cautiously to watch the traffic, the steady flow of automobiles rolling through the snow. On the streets that were paved, they sounded like wooden sleds slicing through an icy hillside. More than the weather had created a sense of chaos on the streets. Streetcars ran along tracks, cars and double-decker buses wove in and out of any free space, and then there were horse-drawn buggies. With no real lanes and no speed limits that could accommodate that many modes of transportation, New York streets were chaotic. The ice only slowed the chaos.

As he crossed the street, Tilney hunched his shoulders and pocketed his hands against the wind gusts. He was clean-shaven, as he always was; Tilney didn't particularly like the trim "facial forestation" that had become so popular. He had a round face, accentuated by round spectacles, and managed to look boyish through most of his life. Like most men, he wore a sacque suit, the one he would wear all morning, with a tie and tiepin. Sometime in the afternoon, Tilney would return to his office and change into an afternoon suit after soot, ash, and a fine film of soft-coal grease had soiled the morning one. On some days, the pollution was so bad that ocean liners had to anchor in the harbor and wait out the impassable yellow fog.

Tilney was on his way to see a new patient that morning, a girl named Ruth. As he neared her neighborhood, he passed men and women along the sidewalks, braving the weather. Victorian wide-brimmed hats brushed by him, the women draping their skirts in one hand, their long hemlines crackling with ice as they walked by. But the younger women modeled the dramatic changes in fashion since the war. Women accustomed to being nurses and wearing pants in the trenches, or working long hours in the war factories or at volunteer stations, had not come home to embrace the corset. Clothing styles were also changing because of their availability; the Garment District was mass-producing simpler styles. Most of the women Tilney passed that morning wore shorter, straighter hemlines and bell-shaped cloche hats. Some of the daring ones had even cut their long hair into a "Castle bob," like popular dancer Irene Castle.

Tilney did not know what kind of woman his patient was that morning—what she was like, how she dressed, how she thought—and he never would. All he knew was that Ruth was a sixteen-year-old who presented with a number of unusual symptoms, the most pressing one being that she was asleep and would not wake. So her parents had sought help from one of New York's best-known and most well-respected neurologists.

Fred Tilney, as he was called by friends, seemed to see medicine differently than other physicians. He was willing to consider the unknown.

In what would become his most famous piece of writing, Tilney would state that most men use only a quarter of the 14 billion cells of the brain cortex and that "the brain of modern man represents some intermediate stage in the ultimate development of the master organ of life." Tilney predicted that "once man is able to use all of his brain cells—maybe a thousand years from now—he'll be wise enough to put an end to wars, depressions, recessions, and allied evils."

In fact, choosing neurology as his specialized field showed an attraction to the unknown. It was relatively new and uncharted. Up to that point, doctors were treating neurotic patients with fever boxes—placing them in heated boxes, hoping to raise the body temperature enough to kill germs. Or patients suffering from hysteria were treated with "crown breezes." They were strapped into something that looked like an electric chair and given continual, small doses of static electricity that literally made their hair stand on end.

In America, neurology first came to the public's attention with the sensational case of Phineas Gage in 1848. The railroad worker survived an accidental explosion that sent an iron rod through his head, entering and shattering his jaw, passing behind his eye, and piercing through the top of his skull. The three-and-a-half-foot rod landed several hundred feet from Gage, and in spite of his wounds, he stood up, walked, and talked. What should have been a fatal accident instead left Gage relatively normal—except for a marked change in his personality. It gave physicians direct clues as to which parts of the brain control which functions.

If Gage's was the first case, the first major wave in neurology came with the Civil War. Only the *term* "shell shock" is unique to World War I; survivors of the Civil War also knew it, calling it "Soldier's Heart." By the 1880s, neurology was taking off—primarily in Europe—and up to that point, America had been lagging behind. It wasn't until the turn of the century that America's interest in "nervous disorders" began to progress. Modern life, it seemed, had put too much strain and stress on the human nervous system—automobile accidents, airplane testing and crashing, the recent anarchy movement, and epidemics of disease, among other things, were taking their toll. And then the Great War filled clinics with glassy-eyed, shell-shocked soldiers or survivors of severe head wounds. For brain researchers, the distinction was a moot point. An injury to the psyche or an injury to the actual brain demanded the attention of the same specialists.

As a testament to that fact, early twentieth-century neurology was actually referred to as *neuropsychiatry.* It was an important term, one that saw the mind and the brain as one entity, but soon that would change. In later decades, the *brain* would be seen as the gray matter surrounding the central nervous system of any animal; the *mind* would be defined as the center of emotions and thought processes distinct to humans. In much the same way, nerves would be disunited from nervous conditions, the former becoming the realm of neurology, the latter the domain of psychiatry. In Tilney's lifetime, those two fields would sever from one another, and never again in medicine would the brain and the mind be part of the same medical specialty.

In the 1920s, the dual force of neuropsychiatry would prove to be essential. When encephalitis lethargica struck, physicians were confronted with a disease of the brain that wholly affected the mind.

Aside from war wounds, the other reason brain study was becoming popular was through the impressive work being done in radiology and image technology. Up to that point, the brain had been the mysterious, unseen part of the body. As one observer noted, "The X-ray threatened to expose the two holiest sanctums of the human body—the sex organs and the brain—and the process demystified both." When Thomas Edison invented the lightbulb, he was just steps away from the X-ray, and he soon attempted early versions of the X-ray machine. His closest assistant, Clarence Dally, operated the X-ray tube—holding the images with his left hand. Dally soon suffered burns and hair loss, then he lost fingers and eventually had to have his left hand amputated. Still the burns and infection spread, until Dally lost both arms and, ultimately, "died by inches." Distraught, Edison abandoned the project altogether. The research on image technology did not stop, however, and a lost finger or amputated hand became the emblem of a radiologist.

By World War I, X-rays were in regular use for identifying brain fractures and finding bullets and foreign objects in the body. X-rays were becoming so commonplace they even became public attractions. Fashionable New York women had X-ray images taken holding hands with their betrothed, and there were coin-operated X-ray machines on the street or in shoe stores, called "Foot-O-Scopes."

Still, it would be several more years before X-ray technology could identify emotions or even brain death, so in the early decades of the twentieth century, X-rays only illuminated the mystery beneath the skull, and neuropsychiatry was the field that delved into that mystery. As a result, neurology was a frightening field full of unknowns in an era that liked certainty.

One reason Tilney may have been willing to choose such an uncertain field of medicine was because he had never intended to be a physician in the first place. He had been born and raised in Brooklyn and went on to attend Yale. At school, he had been editor of the *Yale Literary Magazine*, and in 1895, he contributed a poem he had written, one that would seem to foreshadow his work to come:

> It was thy gold, oh butterfly,
> That caught the childish fancy of my eye,
> But when within my hands thy powdered gold fell off,
> I cast thee by to weep,
> And then again in dreams I'd chase thee in my sleep.

After graduation, Tilney went to work as a cub reporter for the *New York Sun*. At that point, and for reasons he never explained, he decided to become a professor of medicine.

In order to pay his way through medical school at Long Island College of Medicine, Tilney wrote baseball stories for a boys' magazine. His stories became so popular that his editor approached him to stay on full-time, offering him a $5,000 salary. Tilney must have been committed to medicine by that point because becoming a doctor rather than a sports reporter would mean a major pay cut. When Tilney later became president of the American Neurological Association, his annual salary would be only $1,000.

As he had done with his writing, Tilney would attract attention for his talent in medicine. A classmate named Roy Chapman Andrews, who would later become director of the American

Museum of Natural History, noticed the young Tilney near him working intently on their comparative anatomy assignment. The class professor leaned over to Andrews and said: "You watch that fellow. He is the most brilliant student I've ever had in neurology. He'll be a great doctor some day."

Andrews and Tilney were often in the labs, late in the night, at Columbia's College of Physicians and Surgeons, breaking from their studies to eat hamburgers and drink beer in the early hours of morning.

At that time, the study of medicine took a certain amount of fortitude. Prisons had recently adopted electrocution as their solution for inmates on death row. For physicians, it was a chance to study the effects of electrocution on the internal organs, so crisp corpses regularly arrived from Sing Sing.

Tilney was graduated from medical school as the valedictorian of his class and soon left for Berlin to study neurology further. After all, at the turn of the century, Berlin and Vienna were the neurological centers of the world. The labs of Germany had long been superior to most others in the world, and in Vienna, psychiatry as well as neurology was gaining attention. Tilney's work in Germany would leave a lasting impression, the shadow of an idea that would begin to come into focus in the coming decade.

When Tilney returned to New York, he began teaching at Columbia University, and in 1920 he was asked to join the newly formed Neurological Institute. Tilney became part of a major transformation taking place in American medicine. In the previous century, America's medical teaching, considered far inferior to Europe's, focused on the general practitioner and the body as a whole. Its inner workings were based on a mysterious, intricate system of balance. Preserve that balance, and good health could be maintained. Physicians were educated through apprenticeships and vague medical schooling; degrees came too easily. As a result, there were quacks, odd medical practices, and a long list of herbal tonics having very little to do with science. In the twentieth century, at the height of the Progressive Era, medicine evolved into medical specialties. Germs and viruses had been discovered, putting an end to the mystery of what causes disease. Medical schooling became more rigorous and more expensive; degrees were harder to attain. Doctors specialized in particular fields; and, as opposed to days past when they went straight into practice, physicians completed their education by working in the hospitals as interns.

It seemed ridiculous to Tilney that America boasted hospitals all over the country dedicated to the eyes, ears, skin, obstetrics, surgery, the crippled, and the convalescent yet had no hospital devoted to the nervous system. For America, that was still a virgin field. One of the institute's founders wrote, "Future generations will call us pioneers." (Incidentally, Tilney's devotion to specialized schools didn't end with medicine; he once wrote an article in the *Washington Post* explaining the need for schools to train politicians.) But it was a central institute for neurology that Tilney believed in the most, adding, "a brain institute would be of more good to civilization than a whole fleet of battleships." Another time he said, "It is amazing how little interest man has shown in his brain, the most important organ of his body, which controls his work, his happiness and perhaps his salvation."

Tilney also saw an opportunity taking shape in neurology. He had seen the labs in Germany, and he knew they were far superior to anything in America. But he also knew Europe had been waging a war against itself. The continent was cracking under the warring factions, and it was bleeding away the advantages it held. This could be the perfect time for America, and specifically New York, to become the neurological center of the world.

The original Neurological Institute was a mere six stories high and sat sandwiched between a police station and a firehouse on Sixty-seventh Street. Physicians and patients listened to sirens through most of the day and night. The building had one operating room and one operating table with broken legs fastened together by a wooden splint. In the corner of the operating room sat a stand with a cracked water basin, and in another corner was a broken copper water sterilizer. A thin layer of dust coated everything in the room. Still, in the first year alone, fifteen operations on the brain were carried out at the Neurological Institute, another twenty-one surgeries on the spine, and still another twenty-one on the nerves. The institute quickly grew from one room and one technician to occupying the entire floor to requiring a new, state-of-the-art building. Even more impressive was the number of outpatients who flocked to the institute. In its first year of operation, a little over ten thousand came for treatment; by 1912, just two years later, the number had nearly tripled.

Tilney's reputation grew as well, and by 1918, he was president of the New York Neurological Society. He was also teaching at the Neurological Institute. One of his former students described him as a born teacher and an extremely erudite gentleman: "He was a tremendous worker, and he had a combination of qualities which are rare. His keenness of intellect and his power of expression in speech and writing were unusual." If anything, his only drawback was that his good nature and amiability could almost be a weakness. The same student went on to explain that Tilney had no knowledge of the pettiness in life; he was overwhelmingly kind.

At the time, Tilney was a talented, ambitious neurologist. In the next decade, he would become known as the country's greatest specialist on brain function.

Tilney climbed the steps to Ruth's home and rang the bell. He removed his hat, kicked the ice off his boots, pulled his hands free of his gloves, and held tight to a leather medical bag full of instruments and medications.

* * *

In New York, in neurology, in the world, progress carried science and medicine further during the early twentieth century than at any time in the past. A historian of the time period noted that we learned yellow fever was spread by a mosquito, typhus by a louse, bubonic plague by a flea, typhoid and cholera by germs in water and milk. We discovered insulin for diabetes, vaccines for viruses, adrenaline, antitoxin for diphtheria, X-rays for imagery, treatments for syphilis, and radium for cancer, as well as other major advances for surgery and sanitation.

Medical science was easing the burden of mortality rates and cycles of mourning. For parents, there was a sense of relief. Their children would not have to grow up with the fear and curse of epidemics. Losing one or two or three of their children to disease no longer consumed their every thought. But maybe it should have.

CHAPTER 5

Ruth

It took a moment for Tilney to adjust his eyes from the bright light of snow to the dark house, which under coal drafting had extinguished all lights at sunrise. White winter light was coming through the windows and, in spite of the heavy winter drapes, silvering the soot that floated through the room.

Tilney was led from the drawing room to Ruth's bedroom to see the sleeping girl. He did not know for certain what Ruth had, but based on the symptoms her parents described, he had a pretty good idea. Tilney had heard of an epidemic of encephalitis occurring in Europe; doctors were calling it by the technical name "encephalitis lethargica," but the public referred to it as "sleeping sickness."

He had already seen a few cases he believed were sleeping sickness. In fact, he may have seen the first. The year before, a mother had brought her limping son to see Tilney. The boy had become ill while traveling to England with his mother, and because they had come from New York just as the city's largest polio epidemic was dissipating, the English physician they saw had assigned a fairly simple diagnosis. Tilney didn't have access to the case notes, but he wasn't sure the diagnosis fit. The boy had fallen into a deep sleep for fourteen days straight. When he had finally come to, it took several more weeks for him to have the strength to lift his head from the pillow and months before he could fully walk again. His right leg never recovered and soon lost the sculpted shape of a child's muscle. By the time Tilney saw the boy in New York, dragging a lame leg into the office, Tilney believed the illness had been something else entirely. He later wrote, "In no other way did it seem possible to explain the unusual and prolonged somnolence."

The few cases of sleeping sickness Tilney had seen since then were disturbing. In all of the cases, Tilney had noticed inflammation of the brain tissue. What was unusual was the range of symptoms. Some died in a deep sleep, while others died from insomnia—literally, staying awake for days before their bodies gave out. Even more frightening was what English doctors had been reporting: that there was not only physical damage, but mental damage as well. Epidemic encephalitis was literally a kaleidoscope of symptoms. How was it possible to have an epidemic of sleepiness, sleeplessness, paralysis, hyperactivity, and hysteria all at once?

Tilney feared that Ruth's would be one more case of the disease that was building toward an epidemic in New York City, though no one yet called it an epidemic—especially not the health department.

Ruth had been described as a healthy and robust girl, but when she returned from work the week before Christmas, she started complaining of a severe pain in the index finger on her right hand. It was acute and came on suddenly that afternoon.

By the time she reached home that evening, the pain had spread up her entire arm. The arm ached for hours, and then the pain disappeared suddenly, leaving her right arm slightly paralyzed. Up to that point, an aching arm and what seemed like joint pain had been her only symptoms. The symptoms had come on quickly, but certainly did not point to anything more disturbing to come. And so it was both startling and frightening when Ruth flew into a sudden rage.

She became irrational and violent, lashing out at her parents. It was as if their daughter had gone insane—immediately and without warning. The family history presented no tendency toward mental illness, and although Ruth's older sister had epilepsy, no epileptic seizure had ever been like this.

With wild eyes, thrashing limbs, and clenched teeth, Ruth finally had to be sedated and restrained. Then Ruth fell asleep. Even as she slept, her temperature rose to 102. Her parents could only stand back and watch their daughter, strapped to a bed, gasping in stunted, rapid breaths of air like an animal. At first the sleep must have been a relief, but as the days and weeks passed, it would become terrifying.

Ruth's eyes had closed just before Christmas, and they had never opened again. It was now nearly February, almost as if she was hibernating through the harsh winter. There would be no memories of that Christmas, of snow on her windowsill or fire smoke or cold sunsets that turned the sky the color of plums. While she slept, she had lost weeks of her life.

Tilney studied the young woman in the bed. Ruth was being kept alive with a feeding tube and a Murphy drip for hydration. She was still very weak and lethargic; her eyes and mouth were closed. When prompted, she could move her body side to side.

The only thing that took away from the serenity of what appeared to be a sleeping girl was her shape. Her arms were rigid and bent out of the sheets, her fingers flexed. One doctor would describe it as resembling the "effigy on a tomb." It looked like a position impossible to hold for any significant length of time, and yet Ruth had been that way for hours. When Tilney went to the bed to examine her, he pulled gently and found that her arms could be moved out of their statuesque pose.

All through her body, her muscles were taut, her limbs rigid—not the stiffness of something frozen still, but the agonizing stiffness of muscles tensed. And still, even under that physical pressure, her muscles never strained or relaxed.

Like her body, Ruth's face was fixed and masklike. Tilney shined a light into her eyes, but they were nothing more than two dilated spheres of blackness. She showed no reaction to the light. Nor would the muscles around her eyes twitch or wink. She also failed Tilney's tests for smell and sound. Even if she had been able to identify a sound, smell, or sight, her body had no way of responding.

Ruth was fixed in this state, unable to move, and when Tilney tried to move her, she began to tremble. It started with a hand or arm and then spread until her entire body was convulsing in a rabbitlike twitch.

It's remarkable how quickly the body becomes grotesque and crude when disconnected from control of the brain. Ruth's mouth would drop open; saliva would run down her chin. Her hands would clench. She would shudder and shake. She was not paralyzed, at least not in the sense that she'd lost all feeling. She could feel her bed, and she could feel the restraints on her wrists and ankles, the weight of them against her skin.

Surely, too, she smelled the scents that came and went from the room, felt the cold air that rushed into the room whenever a window was opened. She heard the sound of icicles falling from the tree limbs and shattering on the street during the night. And she listened to the voices.

It was Dr. Tilney's voice that told her to blink her eyes or squeeze his hand. His voice asked her whether or not she could smell pungent odors or hear his voice. Of course, she could, but she had no way of telling him so. And Tilney was the one who bent her limbs, and, like a puppet to its master, her body would respond. But then the tremors would begin, shaking her body so hard she could feel the bed beneath her quaking.

To compile his case study, Tilney also inquired about family history, Ruth's eating habits, how many cups of coffee she had a day, if she drank alcohol, if she took exercise. He photographed the girl for his file. He then followed up with further tests—her blood work was normal, the Wasserman test was negative, her spinal fluid was clear. Everything but the girl herself appeared normal. Tilney's case histories covered as much as possible of the patient's life. Since Tilney's handwriting, like most physicians', was hurried and illegible, he often typed up his cases, and his typed files answered the usual questions.

> Character of Birth: Premature or late, breach or complications?
> Status of Birth: Color of infant, weight and size of infant?
> Diseases: All of the usual diseases, blows to the head, malaria, abnormal crying, eccentricities, conduct, poisoning, sleepwalking?

For treatments, he prescribed sponge baths to control fever, high colonics given daily to flush the system of any germs, and frequent lumbar punctures. In cases of insomnia, barbiturates were prescribed.

In Ruth's case, she remained in a deep sleep and could not be roused. Tilney was at a loss—here was a patient who looked to be in a coma, but wasn't. A patient who appeared paralyzed, but could be moved, bent, and changed into different positions like a doll. A patient who showed no response whatsoever to light, sound, touch, or smell, but could react when told to blink or squeeze a hand. Somehow that made it worse. There was not even the comfort that Ruth was unaware or oblivious to what was happening to her. Ruth was truly imprisoned in her own body, a child, palms pressed against the window, looking through the glass.

Finally, standing just beyond Ruth's bedside, Tilney quietly told the girl's parents that there was nothing else to be done. He had performed every possible test; there simply were no answers. Ruth seemed to fall away from their reach, deeper and deeper into her own closed world like a face disappearing beneath the surface of water. Tilney apologized and in quiet tones told Ruth's parents that she would never recover.

When Tilney turned and looked back at the sleeping girl, tears were running down her cheeks.

Tilney walked back to his office in a city draped in the bleakness of a winter dusk. Night was already beginning to fall, bringing the smell of gas from the marigold globes of street lamps. As the dark softly enclosed the buildings, their shapes turned blue, then gray, then black, as they were lost to the darkness. And in this ashen light, he could see frost still lining the lampposts and fog coming off the river, settling over a rice paper moon.

A few days later, as Tilney had known it would, Ruth's temperature rose to 107 degrees, her pulse raced as high as 170 beats per minute, and she died.

CHAPTER 6

The Neurologist

By the time Ruth died, American physicians had been warned to be on the lookout for an unusual form of influenza that was affecting the brain, but because of the war, their information was stunted. It was impossible to track the epidemiology of a pandemic when the world was at war. The only thing European physicians could pass along was dire news: the disease seemed to be changing as it spread. And it was for the worse. Few things could frighten medical investigators more than the knowledge that their microbial opponent was shifting, changing, evolving—especially one in the brain.

Tilney himself had now seen several cases of epidemic encephalitis—some catatonic, some who died of respiratory failure, some who recovered, and nearly all showing signs of brain damage. Tilney was compiling all of his cases to include in a book that would be published the same year. In his book, he described the various cases and included photographs. The photos themselves have that detached, voyeuristic quality of medical book illustrations. In one, a sixteen-year-old girl with long dark braids is standing naked. Her posture is stiff and her fingers flexed. Her eyes are dilated, their glassy gaze staring back at the camera, and she wears a masked expression on her face. Tilney remarked in the text that the girl would sometimes smile uncontrollably.

In another series of photographs, a man's face changes expression through four frames. The photos, like all the others, depict the strange statuesque quality of their subject. Though the photos are still-life portraits of people, the immobility of the patients, their bodily prison, is evident even in black-and-white.

And Tilney's book, of course, included Ruth's case, listed as fatal.

Dr. Frederick Tilney's fame in the world of neurology was also just beginning. In the coming decade, his reputation would reach new heights. He would become a professor of neurology at Columbia's College of Physicians and Surgeons before the age of forty. And he would be the only neurologist ever to treat Helen Keller. She was never seen by a brain specialist as a child, and by adulthood, the public was becoming skeptical. Many thought she was faking her disabilities.

Tilney examined her eyes and found no retina; he tested her hearing and confirmed her to be deaf. Tilney then wanted to test a theory of his—that a person lacking in certain capabilities will compensate with others. He hoped to prove that Helen Keller would be more sensitive than the average person, so he drove her twenty miles or so from her home in Forest Hills to Garden City. Tilney purposefully left the windows rolled down and watched his passenger for several minutes before asking if she could tell him anything about the countryside they were traveling through.

Keller told him that they were making their way through an open field, that they had passed a house with a fire burning, and that there were several large buildings nearby. Tilney confirmed that they were driving through an open field—the road ran along a golf course—and in the distance he could see a cottage with a twine of smoke rising from its chimney. Even the large buildings were there—they were driving near the Creedmoor State Hospital for the Insane. In spite of her astute observations, Tilney was disappointed to discover Helen Keller tested average and showed no extraordinary sense of smell, but Tilney's tests were also limited. Keller herself would say, "In my classification of the senses, smell is a little the ear's inferior, and touch is a great deal the eye's superior."

Tilney continued his friendship with her, exchanging letters and often visiting her to further test her mind. He once placed a coin in her hand and told her: "This was the one touch of nature which made all men kin." Without hesitation, she responded, "Pessimist."

It was his experience with Keller that would lead Tilney to the conclusion that humans were using only a fraction of their potential brainpower, a theory that would become famous in his book *The Brain: From Ape to Man.*

Helen Keller was Tilney's most famous patient, but he treated other prestigious patients as well. Henry P. Davison was a banker and well-known supporter of the Red Cross, and Tilney was treating him for a brain tumor. Tilney once casually remarked to Davison that he would love to study the brain of a porpoise. Davison took his elephant rifle, went out to sea, and brought one back for Tilney. In addition to the trophy porpoise, Tilney studied the brain tissue of apes, rats, reptiles, birds, and fish. One of his more famous animal studies was that of John Daniel, a gorilla in the Ringling Brothers Circus. When the gorilla died, several of the voracious scientists of the Progressive Era wanted to study every part of him. The American Museum of Natural History gave his brain to Tilney.

Tilney's list of human patients was growing as well; one of his most prominent patients was Adolph Ochs, owner of the *New York Times.* Ochs's family placed him under Tilney's care when he showed abrupt periods of happiness and energy, followed by deep depressions, what is known today as bipolar disorder. Another famous patient was Alfred Stieglitz's daughter, Kitty. Stieglitz, the famous New York photographer, was married to Georgia O'Keeffe at the time. Kitty, O'Keeffe's stepdaughter, suffered from debilitating postpartum depression following the birth of her son, and Tilney was chosen as her doctor.

Tilney would also treat two of the best-known victims of epidemic encephalitis during the 1920s. Both cases would lead to significant financial contributions to the study of the disease: one because the patient survived it, the other because the patient did not.

Tilney was becoming one of the country's experts on epidemic encephalitis. But what he did not yet understand about this disease was that death wasn't the worst fate. Dying of encephalitis lethargica would not prove to be the tragedy; surviving it would.

CHAPTER 7

The Medical Investigators

On February 4, 1919, Tilney prepared for the address he would give as the retiring president of the New York Neurological Society. As he did so, encephalitis lethargica, the disease the medical community was still loath to call epidemic, was spreading rapidly. In New York, health officials bickered over whether or not the disease was accurately diagnosed, much less epidemic. So far, the disease had not even been mentioned in the newspapers. The first article would appear in the *New York Times* in March of that year with a warning from the health department about a new, unexplained disease from Europe beginning to spread.

The war had provided the first opportunity encephalitis lethargica had to crawl across the world with little notice from the medical community. And by 1918, the pandemic flu had given it the second opportunity, stealing worldwide attention, infecting and killing millions. Epidemic encephalitis moved with the flu, almost like a parasite to a host, often attacking many of the same victims, receiving very little notice at all.

By the time Tilney retired from the New York Neurological Society and prepared his closing remarks, the city of New York had lost well over 100,000 people to the pandemic flu. Doctors had fought a losing battle against influenza, but Tilney and other neurologists saw opportunity in encephalitis lethargica. A fellow neurologist in New York wrote about the epidemic, "There has been no occurrence in the field of neurology that has been as illuminating as encephalitis, in that its manifestations may ape practically every neurological entity."

With more cases appearing each week, and Tilney's tireless work at improving the Department of Neurology, he was more convinced than ever that New York itself was positioned to lead the world in encephalitis research. Europe was still in chaos after the war, and all of the destruction had slowed intellectual pursuits, narrowing Europe's scientific lead over America. What's more, World War I had made the United States a world power, and the flu pandemic had made American medicine a respected force of both public health control and laboratory research. That was precisely what Tilney chose to tell the members of the New York Neurological Society: "This fact should be recognized at once in America as an opportunity for service, the purity of motive being enhanced by the lack of aggrandizing competition."

Tilney proposed to coordinate all of the neuropsychiatrists in the New York area, and as one medical historian wrote, "This coordination of neurological research . . . quickly coalesced around epidemic encephalitis." Cases of encephalitis lethargica were now appearing not only in Austria, France, and England, but in Russia, Italy, and parts of Africa and Asia. It was as if no corner of the world remained untouched by this strange disease. In the United States, cases were surfacing in major cities all over the East, extending as far south as Texas and as far west as Illinois, and the number of patients was about to triple. But it was New York that saw the highest number of cases, and it was New York that would become the focal point of research.

Another factor made the city ideal to sit at the helm of American medicine. Unlike in many other cities, New York's medical community had a close association with its public health institutions. Epidemiologists would have little chance against this disease without the tireless efforts of New York's Department of Health.

Standing outside the door of a dank and dreary tenement house to take away an infant with polio, tacking the quarantine notice to a building where a smallpox case lived, sending a tuberculosis patient away to a sanitarium—the work of the public health department of New York City was challenging to say the least. One year, the department removed more than 17,000 dead horses, mules, and cattle from the city; it carried away almost 300,000 smaller animals inside the city; and it inspected nearly 4 million pounds of rancid poultry, fish, pork, or beef. Another year, the *New York Times* published a list of the dead animals that the health department removed from within the city limits: 3 elephants, nearly 6,000 horses, 308 cattle, 11 colts, 16 ponies, 1 alligator, 2 camels, 1 bear, 1 lion, some 50,000 cats from streets or shelters, and even more dogs.

Before plumbing, the health department also handled privies and raw sewage, as well as keeping the Croton aqueduct a clean source of water. Since its inception in 1866 until well into the twentieth century, filth, more than disease, had been the department's main concern. In the past, the health department could do little more than issue sanitation orders, track disease into certain neighborhoods, and quarantine areas. But that was changing.

Measures as small as coating open water sources with petroleum to kill mosquitoes or providing public bathhouses for the poor or requiring laundry trucks to keep clean towels in public restrooms fell under the jurisdiction of city health. The health department performed tasks as small as inspecting hot dog stands and rounding up drug addicts on the street, but it also had the power for larger initiatives like closing businesses, banning public funerals during an epidemic, or forcing schools to provide open-air classrooms for their tuberculosis patients. The department went so far as to prohibit smoking in certain places—more to address the nuisance of the smoke than the ill effects on health. If smoking was not controlled, health officials warned, there might be a day when smoking would be prohibited in *all* public places in New York.

Since the 1890s, New York's health department, the best in the world, had battled politics as much as microbial foes and lapses in sanitation. The department was trying to maintain a safe distance from the grasp of Tammany Hall, the Democratic political and social machine that had dominated New York politics for nearly 150 years as a force of reform by way of corruption. Tammany politicians wanted control of all municipal departments, among them public health. But rather than filling the department with qualified physicians and scientists, they gave positions out to loyal supporters, whether or not they were qualified, whether or not they even held an M.D. As a result, several doctors serving in the department resigned and announced publicly that they did so because of Tammany control. In other words, public health was so important, so vital to the city's progress, that it superseded politics.

Epidemics of cholera and diphtheria in the 1890s put the health department in a role of greater importance and gave recognition to its newly formed Division of Pathology, Bacteriology and Disinfection. The bacteriological laboratory would thrust New York into the spotlight of world medicine and public health. That was a remarkable feat—there were the Johns Hopkins labs, Mayo Clinic, Pasteur Institute in Paris, Lister Institute in London, the clinics in Berlin and Vienna . . . and New York's municipal health department lab.

Out of that laboratory, and under the talented direction of Hermann Biggs, and later William Park, the diphtheria antitoxin would be developed, as well as groundbreaking work on pneumonia and meningitis. The labs and the New York City Department of Public Health became so valuable that as Tammany again tried to gain control in 1918, in the midst of the great flu pandemic, the U.S. government stepped in and forced Tammany to a sudden halt.

The health department's future success would depend on separating itself from political strings and adherence to the way things had been done in the past, and instead devoting its funds to a more modern approach to medicine and the study of disease. New knowledge about germs and viruses enabled health officials to take a direct approach in controlling the spread of disease, to become active, not just *reactive*. Their approach was becoming so aggressive, in fact, that it often stigmatized certain ethnic groups and threatened civil rights. That fact would be obvious on Ellis Island. At the turn of the century, roughly 2 percent of immigrants were turned away due to disease; by 1916, it was up to 69 percent.

Mary Mallon's story is one of the most famous examples of the complicated relationship between civil liberties and public health. During an outbreak of typhoid, an engineer was asked to examine the aqueduct and other water sources in the city to find contamination. As it turned out, the "contaminant" was a person, a carrier of the disease. "Typhoid Mary," an Irish cook for some wealthy New York families, did not have a case of typhoid, but she carried the germs nonetheless and spread them in food. The health department demanded that she stop working as a cook. Wary of the health officials, Mary moved from job to job as a cook, each time leaving several people sick or dead of typhoid. Finally, the health department imprisoned her in an isolation hospital for the rest of her life.

When it came to infectious diseases, community health overruled individual rights. Physicians as well as citizens had a legal responsibility to report any cases of infectious disease; individuals could be isolated and quarantined in their homes; even landlords were responsible for not allowing a sick tenant to leave or move without consent of the health department.

Aggressive disease prevention became the full-time commitment of the New York City Department of Public Health. The department focused on educating the public: teaching better hygiene and preventative medicine, keeping neighborhoods clean, coordinating reports among hospitals, sending out informative pamphlets (in several languages), and keeping ever mindful of the fact that if disease was not directly a product of its environment, the environment certainly played a part in the spread of an epidemic.

The health department was called to action during both the 1916 polio outbreak and again in the 1918 pandemic flu. Polio had forced public health workers into a frenzy of preventative measures. They quarantined homes, physically took infants and children to protected wards, and even banned children under the age of sixteen from public places like theaters. At first the health department focused its efforts on the immigrant neighborhoods, but it quickly became obvious that polio was not attacking the Irish or Italian tenements; the disease was striking wealthier children in the cleanest neighborhoods. Polio is one instance when progress worked against civilization: the virus is spread through filthy water, and when water systems were cleaned and made safer, people lost their immunity to this pathogen that had regularly circulated in soiled water for centuries. That's why polio largely became a twentieth-century plague. In spite of the health department's best efforts, by summer's end, polio left thousands of maimed survivors, most of them children, in New York City.

Interestingly, the 1916 outbreak of polio afflicted nine thousand people and went down in history as the most devastating polio epidemic in New York. By the time epidemic encephalitis would suddenly and inexplicably disappear, it would infect at least five thousand New Yorkers, and it would not go down in history as anything at all.

* * *

With barely a reprieve after the polio epidemic, and in the midst of a world war, influenza hit the city. The health department again rushed to action. It published notices prohibiting public spitting, and it closed stores, except those selling food or drugs. It even staggered work hours to avoid concentrations of people on mass transportation during rush hours. Even so, the epidemic became so overwhelming, with more losses than the war's casualty lists, that the department had to intercede in areas it would not normally oversee—food distribution to the sick, converting private homes into makeshift hospitals, even caring for children suddenly left without parents.

New York City and its public health department had been dealt a series of blows. Still, in 1918, the *New York Times* reported that "the Health Department of the City of New York stands as a model of municipal health service in this country, and has been recognized abroad."

In the mid-nineteenth century, New York had the highest death rate of any American city, higher than London or Paris for that matter. By the end of World War I, New York City's handling of public health was considered exemplary. In everyday life, it could be a nuisance, but when a major epidemic struck, the public depended entirely on the health department.

Often, it is not how a city functions on a daily basis that is the measure of its civic success, but the way it survives a tragedy. New York would rise to that occasion time and again.

Tilney was not the only one to see New York as a rising star in worldwide medical research. Five months after Tilney's address to the New York Neurological Society, Royal Copeland, the city's health commissioner, continued work on a plan to place New York as a medical center that rivaled Vienna and Berlin. Copeland had been planning to retire, and as one newspaper account tells it, Mayor John Hylan held his fist to Copeland's face and demanded he stay in the position out of patriotic duty. Copeland spearheaded a nationwide campaign to raise $30 million for public health and medical education. He wanted New York to offer the same valuable education that drew American physicians to Berlin, Vienna, or Paris for study. After all, he argued, there were more hospital beds on Blackwell's Island than in all of Vienna. Furthermore, Copeland argued, there were places in the world, like India, where the average life span was as low as twenty-one years; in New York, it was as high as fifty years. The bacteriological lab of the New York health department, Copeland added, was the greatest vaccine laboratory in the country. America, which had never been able to compete against a European medical education, was now being recognized worldwide for its exemplary public health.

New York's research lab was a critical part of that public health success because it allowed the doctors to grow cultures and diagnose a disease based on science and not a set of vague symptoms. It also turned the focus of public health into a full-time commitment, not one that simply rallied together during an epidemic of disease. During the lull between major epidemic outbreaks, the laboratory would focus on endemic diseases like diphtheria.

In addition to the health department lab, New York's medical community included the New York Academy of Medicine, which itself had been created in part to monitor the health of the city and protect citizens from incompetent physicians. New York's Public Health Committee, under the helm of the academy, gathered and interpreted information from local hospitals. When neurologists began asking for reports of epidemic encephalitis from hospitals, the response was immediate: more than 213 cases in Mount Sinai, Jewish of Brooklyn, and Bellevue. Physicians on the committee were convinced that encephalitis lethargica was in fact its own disease, distinct from flu or polio or any other ailment, and that it was occurring in epidemic form. The health department then made the disease reportable.

Likewise, the Rockefeller Institute, under the direction of medical giant Simon Flexner, had also vowed to discover the germ spreading encephalitis lethargica. In fact, Flexner, having seen many cases of the disease where sleep or lethargy was not the primary symptom, had been the one to publicly change the name to "*epidemic encephalitis*."

All of these institutions were coming together, drawn to the vortex of New York's public health activities. Epidemic encephalitis would give neurology its chance to join the current. The organized and well-run public health offices were thus able to supply the Neurological Institute with an epidemiological gold mine—patient records, statistics, outbreaks. There could not be a better system for investigating an outbreak of disease than to have all of those separate pieces working together. The health department went into the neighborhoods and recorded cases; the public health laboratories diagnosed cases and studied tissue samples; the New York Academy of Medicine provided an endless source of articles published throughout the world; and the Neurological Institute had the funding, space, and talent to conduct brain studies. Certainly, with that kind of combined effort, doctors hoped the mystery of this disease could be solved in the coming decade.

Sigmund Freud wrote in his book *Mass Psychology* that "One of the few pleasing and uplifting impressions furnished by the human race is when, faced with an elemental disaster, it forgets its cultural muddle-headedness and all its internal problems and enmities and recalls the great common task of preserving itself against the superior might of nature."

CASE HISTORY THREE

New York City, 1922–27

NAME: Adam

PHYSICIAN: Dr. S. E. Jelliffe

CHAPTER 8

Adam

It was Easter break in 1922, in mid-April, when Adam left his preparatory school for home. As he traveled on the train through upstate New York, spring was emerging from winter's icy grip. The green shoots of daffodils speared through the dead underbrush. Buds, tiny and jewel-toned, appeared almost miraculously along the gangly branches of cherry trees and dogwoods. And vivid yellow bloomed along the bowing arcs of forsythia branches. Though it happened without fail every year, it always seemed a surprise to see the first trace of life after the long, dead winter.

For Adam, the images of spring were blurred. He shivered with fever and peered through the train window with red, tired eyes. Gusts of wind blew through the window cracks when the train rattled across a rough patch of track. Each breath of air felt like gravel in his throat—burning, raw, and painful. He was glad to have the chance to go home and rest.

Adam's mother put him in his bed, his childhood bed, and he

curled up still feeling feverish and very depressed. The family doctor diagnosed a light case of the flu and told his mother that he should be fine after a few days in bed. The doctor recommended an alkaline gargle for the boy to use every few hours and left a prescription for a cough compound that contained codeine—it would help with the pain and allow the boy to sleep better. The family had no reason to assume it was anything worse than a case of the flu.

The next morning, a little after nine, Adam's older brother, a reporter who worked for a New York daily and still lived at home, came into the room to check on Adam. The room was dark as he cracked the door open and looked at the motionless shape beneath the blankets.

"Get out and let me alone! Let me rest. I had an awful night," Adam shouted.

The brother pulled the door shut, a little surprised by his brother's angry outburst. It seemed out of character.

Later that afternoon, Adam got out of bed and started to wash and shave. His brother watched him through the open door and later remarked that it was obvious something had happened—Adam twitched and jerked and acted as though he labored under some tremendous excitement that he could not control. He was not the same boy who had climbed off the train platform, feverish, depressed, and tired.

Adam was shouting and laughing in the bathroom, singing to himself. When he noticed his brother standing in the doorway, he grinned and shouted, "Wow! I like my liquor strong and my women weak."

Adam's eyes were bright, and his hands moved spastically for his comb and razor. His teeth chattered uncontrollably, and he shook all over. Adam's brother stood in the bathroom, frightened and concerned. It took nearly half an hour for Adam to calm down and for the burst of delirium to subside. His teeth stopped chattering; his hands quit jerking; his body stopped convulsing. Adam looked up at his brother and said, "I can't tell you what happened to me last night except that it was something terrible. I suffered all night. I dreamt that I died and then came to life again and saw angels." They could have been the delusions of a feverish boy, except that Adam's fever was never that high.

Finally, Adam told his brother, "It was just as if I died and came back to life. Mark what I tell you. This is going to change my whole life. I'll never be the same again after what happened last night." He repeated it over and over: "I'll never be the same again. . . . I'll never be the same again. . . . I'll never be the same again."

At first Adam's parents worried that their son's reaction could have been caused by an accidental overdose of codeine. But there were only trace amounts in the cough medicine, barely enough to cause any reaction at all. Although Adam's family was worried, they went ahead and let the boy return to school the next day and hoped that the strange reaction, whatever it was that caused their son to depart so dramatically from his senses and personality, was over.

It was not until summer break that the next episode began. This time, it arrived in the form of a respiratory tic, a compulsive sniffing. Adam also began to talk incessantly and excitedly—again, as though it was something compulsive he could not control. His brother slept in the room with him and was kept awake at nights listening to the quick, unceasing, bizarre sniffing. When the sniffing wasn't keeping Adam's brother awake, his talkativeness was. He asked questions endlessly, coherent questions, but unceasing. Finally, his brother would tell him to shut up and go to sleep. The talking would stop, but moments later, the sniffing would begin.

For Adam, it was maddening. He would lock himself in the bathroom, turn on the faucets, and run water in the cast-iron basin to try to keep from waking his family members with the sniffing. Still, he could not stop the tic. It was like wheezing with asthma or suffering from breathing spasms. The sniffing would stop only if something distracted Adam; if his attention was diverted, the attack would come to a halt.

As strange and random as his symptoms seem, they actually made sense within Adam's brain. The brain is an intricately wired communication system where signals run along pathways to other parts of the brain, almost the way a telephone signal runs along telephone lines. And like a person-to-person call, those signals deliver a message to receptors in the brain telling them to perform a movement or to start thinking or to begin feeling certain emotions. The messages are sent through a system, a switchboard, the same way a telephone operator will connect one voice to another. In the brain, that switchboard is called the basal ganglia, and they connect messages within the brain and to the body, including the nerves that ultimately control movement. The basal ganglia also connect messages going to and from the frontal lobe, the part of the brain that controls personality, behavior, inhibitions, and emotions. In the brains of encephalitis lethargica patients, the basal ganglia are damaged.

The basal ganglia also serve another important purpose: they send messages to the thalamus and neighboring hypothalamus, the small part of the brain von Economo identified as the electrical switch for sleep or the lack thereof. Even today researchers can't explain the exact relationship between those two parts of the brain, but they do know the basal ganglia send messages to the hypothalamus to *halt* certain movements so others can take place. Even with something as simple as falling asleep, the brain is put to work, telling the arms and legs *not* to move, relaxing muscles, keeping eyelids closed, slowing the beating heart and breathing lungs.

In a damaged brain, as messages rapidly travel through the basal ganglia and thalamus to trigger movement, emotion, thinking, or sleeping, they pass through the mangled switchboard and are redirected, sending garbled messages to the body. Disorders associated with the basal ganglia are Parkinsonism, Tourette's syndrome, attention deficit disorder, obsessive-compulsive disorder, cerebral palsy, and stuttering, among others. Whenever these signals, carried by chemicals called neurotransmitters, are interrupted, receptors in the brain waiting to receive the signals get mixed messages instead.

In Parkinsonism, for example, the cells that produce dopamine begin to die off and mixed messages overwhelm the muscles' nerves that control movement. With static messages coming through, the body may have trouble moving at all or lose the ability to stop unwanted movements. In cases of obsessive-compulsive disorder, the basal ganglia forward mixed messages back and forth to the frontal lobe. In an OCD patient, those messages are sent too quickly and too frequently. When that happens, the message overrides logic and tells the person to keep doing the same action again and again. An OCD patient could no more keep from the obsessive thoughts or compulsive behaviors than a Parkinson's patient could keep from shaking, because it is the wiring in the brain automatically controlling the impulse, not their own willpower. Both Parkinsonism and OCD, among many others, were conditions common in encephalitis lethargica patients.

The messaging center of Adam's mind had been damaged when he first contracted encephalitis lethargica. The physicians and his family did not even realize it was a case of encephalitis—they saw symptoms of flu or a bad cold and diagnosed it as such. Adam, like so many other people, would learn only later that his brain had been harmed during his illness. Messages going through the basal ganglia in Adam's brain became rewired, and actions that he normally did automatically, without thinking, were happening out of his control—sniffing over and over again, shaking, stuttering, and other odd movements. The "stop and go" part of his brain would not stop these behaviors. But it was the messages going to and from the frontal lobe that would make the dramatic changes in Adam's personality.

Adam was taken to see several doctors. One found an obstruction in Adam's nasal passage, and surgery was planned, but then

Adam was involved in a car accident and the surgery was postponed. That was in August 1922, and Adam soon left again for school. But it was apparent something was wrong with the boy. He could not pay attention and fell asleep at his desk. He was accused of being lazy and lacking enthusiasm. Adam felt tired all of the time, but was rarely able to fall asleep before 2:00 A.M. Teachers threatened that if he would not stay awake, he would have to leave. He managed to finish the school year, but failed two of his subjects. To those who knew him, it was clear a distinct change had happened. Adam had always been sociable, easy to get along with, funny, a boy who loved music and playing pranks. The Adam that left school at the end of that school year was none of those things.

It had been over a year since Adam's case of the flu, if that's what it was. By the summer of 1923, new symptoms began to arise. The breathing spells became rapid and painful. His face would grow stiff, his hands and feet stiffen, and he began drooling. His brother told Adam it looked like he was foaming at the mouth, and his mother reminded him repeatedly to use his handkerchief. His feet began to shuffle, and he walked strangely. As Adam's condition grew worse and more inexplicable, he was taken to an institution for treatment, where he continued his downward spiral, often refusing to get out of bed until a nurse threw cold water on him.

Adam had been suffering for two years from this strange disease, frequently falling into a trance, panting, and salivating like someone possessed. He had been hospitalized several times, had his tonsils removed, been sent to boys' camp, lived with a physician for a spell, stayed at a farm with a male nurse for a while, and seen numerous doctors. As a last resort, Adam was finally taken to see one of New York's premier psychoanalysts: Smith Ely Jelliffe.

CHAPTER 9

Smith Ely Jelliffe

Smith Ely Jelliffe had an idyllic childhood in a city home with a clear view of Manhattan. Like Tilney, he grew up in Brooklyn, and most of his memories are of his family's brownstone in Park Slope, which, like many houses in the neighborhood, had just been built. In the vacant lots where homes were yet to be constructed, Jelliffe played baseball with his brother, who was only one year younger, and neighborhood kids. Having free roam of the neighborhood, the boys also played "catch one, catch all," climbed the pear and cherry trees lining the streets to steal fruit, and swiped the noonday milk left for the corner grocery. There were croquet games in Prospect Park, as well as kites, marbles, and tops. Jelliffe was the kind of boy who rushed through all of his studies on Friday so he could leave Saturday and Sunday free for play. As with the other neighborhood children, only nightfall brought him home.

Jelliffe had few negative remembrances from childhood. Although he could outrun almost anyone, his arms were weak, so he avoided fisticuffs. And, ever the psychoanalyst, he remembered a distinct sadness one Christmas when he awoke to find his stocking empty on the mantel as punishment for hitting his brother on the head with a hammer. "On several occasions in my later life when I have done some mean or unworthy action . . . this same mood of deep sadness has come over me."

Jelliffe's father was a renowned schoolteacher, the first to organize a kindergarten program in Brooklyn. His mother was an intelligent and vivacious woman who, even at the age of seventy, visited Egypt to ride a camel and see the pyramids; she also traveled through the Panama Canal the year it opened. And she had a wonderful sense of humor that her son inherited.

Jelliffe was also a gifted student. His kindergarten teacher showed up at his doorstep in tears one afternoon, only to tell Mrs. Jelliffe that her son had finished the whole year's work in the first two weeks, and there was nothing left to teach him. Jelliffe attended the public schools in Brooklyn, and at age sixteen he met and fell in love with the girl who was to become his wife.

Although Jelliffe obtained a certificate to teach school like his father, graduated from the Brooklyn Collegiate and Polytechnic Institute, and studied botany at the New York College of Pharmacy, he chose to attend medical school, graduating from Columbia University's College of Physicians and Surgeons in 1889. His schooling seems to have been indicative of not only his wide variety of interests, but his difficulty in choosing one path. In medical school, Jelliffe was drawn to several subjects except one: bacteriology. He recalled that every day during his 11:00 A.M. class, he suffered a migraine each time his professor of surgery said the words "healthy laudable pus." Throughout his career, Jelliffe would find fault with surgery as a cure for disease, believing it to be an "effort to cut ideas out of the body."

After graduating from medical school, at the age of twenty-three, Jelliffe traveled to Europe. As passports did not yet use photos,

Jelliffe's provides a detailed description of his appearance at the time: five feet, nine inches, wide forehead, light blue eyes, ordinary nose, small arched lips, square chin, oval face, light brown hair, freckles. Jelliffe spent a year studying in various clinics in Vienna and in Berlin, even visiting Koch's laboratory and stopping in Paris to hear lectures from Jean-Martin Charcot, the most renowned neurologist in Europe. Jelliffe was particularly interested in the broad availability of knowledge in Europe, where he studied pediatrics, ear diseases, eye diseases, gynecology, internal medicine, and even botany. But he expressed no interest in either neurology or psychiatry at that point. Jelliffe was taking longer than most young men to find his specialty and seemed entrenched in "finding himself."

Meanwhile, an epidemic of influenza was forming in Europe. Jelliffe would call the 1890 flu epidemic his "baptism" into medical practice. In its aftermath, there seemed to be a strange number of illnesses splintering from the flu—epidemic headache, a sort of influenzal meningitis, unusual cases of encephalitis, and oddest of all, a sleeping sickness. It was called nona.

Jelliffe was frustrated with the medical community and its dismissive diagnoses when it came to this disease. He criticized the tendency among the population, as well as the physicians, to "name a peculiar, bizarre, and noisily inconsistent set of symptoms, especially when occurring in women, as hysteria." Jelliffe continued, "To stick pins in an individual and when he does not feel it, or gives no evidence of feeling it, and then say—hysteria—is bumble puppy and not diagnosis."

There was one area in which he showed no indecision whatsoever: his future wife. He wrote letters to her during his year away and quoted poetry. He also encouraged her to follow her own aspirations, which were in science. "It would look fine," he wrote to her, "for a doctor's wife to be a scientist also."

After his return from Europe, Jelliffe hung a sign on his parents' home in Park Slope and started his medical practice. Not surprisingly, patients did not flock to the steps of the cheerful brownstone. But when a cholera scare and a smallpox epidemic hit, Jelliffe found work with the board of health. He also earned money as a medical examiner for a life insurance company.

Still trying to choose a medical specialty and find a viable income, Jelliffe married his childhood love, Helena Dewey Leeming. They would be married twenty years and have five children. They left Brooklyn for good and moved to West Seventy-first Street in Manhattan to flee the "aid of a considerate father-in-law." In the coming years, Jelliffe taught medicine, and he started writing and editing medical articles with the help of his wife. His first published piece was an article on botany, "A List of the Plants of Prospect Park." The move across the river, as Jelliffe called it, also changed him personally. He saw it as a physical separation from his mother, from the comfort and security of his childhood home to a place of independence, even danger, and it was a powerful feeling. The intense feelings made him realize that he knew very little about the mind, and he wanted to learn more.

In retrospect, Jelliffe could see many instances in his life that led him to neuropsychiatry. He had long seen a connection between the health of the mind and the health of the body. "There was no antithesis of body and mind," wrote Jelliffe. "They were one and inseparable but, above all, function determined structure and then structure directed function."

Jelliffe's mind, which at times seemed indecisive and noncommittal, was sharpening into true innovation. His ideas were new age, and his philosophies often controversial. He was known to launch into lengthy, philosophical debates and earned the nickname "Windy Jelliffe" among his students. He was also known to be a fierce opponent—not because he was aggressive or competitive, but because he was so knowledgeable. He had a vast knowledge of medicine and science, as well as a remarkable memory, which, as one biographer wrote, "was more than a match for the average critic." As Jelliffe's career ascended, he continued his own practice, taught medical students, remained active in the American Neurological Association, served as president of the New York Neurological Society, and made frequent study trips to Europe to polish his professional knowledge. He corresponded steadily with his acquaintances abroad. A biographer wrote that Jelliffe became the "conduit through which European medical innovations in nervous and mental disease specialties came to the United States." Jelliffe would be remembered in history as the father of psychosomatic medicine and a pioneer of psychoanalysis.

Jelliffe also served as an alienist. He testified in the famous society case of Harry K. Thaw, who was tried for the murder of Stanford White, a story made famous in modern times by the novel, musical, and film *Ragtime*. Thaw, wealthy and privileged, was married to a beautiful chorus girl named Evelyn Nesbit. When he learned of a past liaison between his wife and famed New York architect Stanford White, he became enraged. In 1906, in the shadows of the Moorish tower atop Madison Square Garden, a building White himself had designed, Thaw fired three shots at close range into White's face. Jelliffe testified for Thaw's team, and in 1908, Harry Thaw was found innocent by reason of insanity. Jelliffe later had to sue Thaw's legal team in order to receive payment for his services.

He also testified in the era's most gruesome serial killer trial, against Brooklyn's Albert Fish, who kidnapped, assaulted, and cannibalized a number of children in the area.

Jelliffe is most famous, however, for his medical writing and editing. He wrote for a number of journals and served as an associate editor for the *New York Medical Journal*. Eventually, Jelliffe found his calling in neuropsychiatry, becoming editor of the *Journal of Nervous and Mental Disease* and remaining in that position for forty-two years. There were a large number of neurological studies coming out of Europe just before World War I, but all were in need of translation. Jelliffe was able to translate and publish the most recent and groundbreaking studies from the labs in Germany, Austria, and France. Jelliffe also helped found the *Psychoanalytic Review*, which first appeared in 1913. It was through the *Psychoanalytic Review* that Jelliffe first became acquainted with Sigmund Freud.

Jelliffe's journey in becoming a Freudian was not a quick one. When first introduced to Carl Jung's work, and later Freud's, just after the turn of the century, Jelliffe quipped, "This whole Freud business is done to death. The lamp posts of Vienna will cast forth sexual rays pretty soon." Children's fairy tales, Jelliffe lamented, would be unsafe material, saturated with sexual symbolism and innuendo. Psychoanalysis as a profession had been launched in the late nineteenth century in Europe, primarily Vienna. Although Carl Jung and Sigmund Freud, both of whom Jelliffe would later meet and correspond with, had gained worldwide popularity and recognition, they had not yet received much respect in the United States. In large part, this may have been due to the more puritanical values of America and the blatantly sexual focus of psychotherapy.

Over the course of several years and long walks through Central Park with a friend and psychoanalyst, Jelliffe came around to the idea. It is not surprising given Jelliffe's intellectual makeup and the strong influence of European medicine on him. He had always been an innovative, open-minded student in search of greater connections in life. He likened psychoanalysis to a surgery of the mind—minor in some cases, deep cuts in others. Like fellow neurologist Fred Tilney, Jelliffe believed that very little of the mind had been explored up to that point. "When one seriously gives himself to the reflection that the human organism . . . has been a billion years in the making, no one but a consummate ass could believe himself capable of understanding but the most insignificant part of this whole evolutionary product," wrote Jelliffe.

Both Tilney and Jelliffe worked at the Neurological Institute during the war teaching medical officers. And, for a time, Jelliffe was also a visiting physician for the institute, working in the outpatient clinic. Eventually, he was asked by a friend to drop out for the sake of harmony at the institute. Apparently Jelliffe's ideas, especially the ones focusing on psychoanalysis, were not always warmly received. At that time, the practice of psychoanalysis was still being "violently attacked" by a number of New York neurologists.

In a sense, Tilney and Jelliffe represented what was happening on a larger scale to the practice of neuropsychiatry. Without seeming to choose sides, Tilney became heavily involved with the Neurological Institute. Jelliffe, on the other hand, saw brain study as more integrated with psychiatry and migrated toward psychoanalysis.

One writer would later capture what seems to encapsulate Jelliffe and separate him from other physicians: "There are those who prefer to gaze backward into the past to ascertain what man has been and the path he has taken to reach his present stage of evolution. There are those who love to analyze the present to learn what man is, and there are others who attempt to penetrate the misty future to anticipate what he will become. Dr. Jelliffe was an amalgamation of all three. . . . He was truly a child of the mists, but practically, he lived for all life was worth and enjoyed it thoroughly."

Just as Jelliffe was hitting his stride in psychoanalysis, he was knocked by the first of two severe blows in his life. In 1916, his wife and childhood love died of a cerebral hemorrhage. She had been his wife and companion, but also a fellow science writer and a translator of many of the articles Jelliffe published. In a book dedicated to her, Jelliffe recognized her "lofty purpose, ideal striving . . . a constant stimulus to progressive endeavor."

Jelliffe was remarried in 1917, to a nurse, but they never had any of their own children. His second wife, like his first, was professionally minded and drawn toward a literary career.

During those years at the height of his career, Jelliffe was a relentless worker. He put in twelve hours a day without lunch. Jelliffe had a remarkable memory and an ever-expanding library of roughly twelve thousand books. He moved his office and home to a brownstone on West Fifty-sixth Street between Fifth and Sixth Avenues. Every room in the home was lined with books, and his office, beneath bright window light, held a microscope and glass jars in a neat row. In his office, his shelves were stuffed full of medical journals. He saw several patients a day, worked on his medical journals, wrote textbooks, then went upstairs to change into a tailcoat, a white wing-collared shirt, and a top hat to spend an evening out at the theater or at dinner with his wife.

On nights that he stayed home, Jelliffe chose the meal—he considered the chemistry of food to be a science. Family dinners were like rituals. They would sit around the dinner table and converse in different languages; one week it would be French, the next week German. There was a small matchbox in the center of the table, and every time someone made a mistake or spoke English, a coin was deposited. His children spoke seven different languages by the time they were adults. And for Jelliffe's children, learning was active, not passive. When he wanted to teach one of his daughters about the Indians native to North America, he took her west to the reservations and taught her to speak some of the tribal languages. Another daughter loved acting, so he helped her produce plays at their summer home, perching in a tree to pull the curtain open and closed. Later, he arranged to have her as an extra in a few Broadway plays with John and Lionel Barrymore. All five of his children were talented and interested in different pastimes, whether acting, boating, swimming, fishing, or gardening.

The Jelliffe household was usually filled with people—family, patients, or the occasional boarder. Whenever possible, Jelliffe escaped to his summer home on Lake George. There, too, the den held books from the floor to its vaulted beam ceiling. The den also held large glass cases filled with stuffed birds, and along the lake, the Jelliffes had an orchard and beehives for honey. The home had been converted from a boathouse, and it was a sanctuary for Jelliffe. Despite the intellectual atmosphere surrounding his homes, Jelliffe's role as a father was anything but austere. He was known to have a sweet tooth, often hiding candy in his office drawers that his children could find. And at Christmastime, he would climb to the rooftop, hanging on to brick blocks and smoke pipes, pretending to be Santa and calling down the chimney to answer questions from the children. One daughter later commented that she never once heard her father raise his voice in anger.

As Jelliffe grew older, his appearance evolved into that of a tall, portly figure with skin as smooth as modeling clay. Several people remarked on his intelligent eyes. Another person described him as something of a Roman Catholic priest, saying, "He was commanding, quizzical, sure of himself, and not to be moved." For his patients, his self-assured presence must have been a comfort, and he had an impressive client list that included Greenwich Village muse Mabel Dodge; several members of the Algonquin Round Table; Betty Compton, Mayor James J. Walker's mistress; and playwright Eugene O'Neill; as well as famous friends like John Barrymore. In fact, when Barrymore played in a controversial, Freudian version of *Hamlet* (making full use of the Oedipal complex), Jelliffe was the consultant.

Other patients, however, were not among the celebrity set or wealthy socialites. True to Jelliffe's nature, but not necessarily mainstream for psychoanalysts, Jelliffe became especially interested in treating psychosis. He embraced the challenge and thought of it as a chess game of sorts. He would sit, patiently and thoughtfully, to watch a mute catatonic for an hour. There was nothing hurried or impatient about his practice, and it drew a large and popular audience. That fall, a boy was brought to see Jelliffe. He had been diagnosed as a hopeless schizophrenic destined to be institutionalized for the rest of his life. His name was Adam.

CHAPTER 10

The Alienist

It was the end of October in 1924 when Adam first visited Jelliffe, and a brilliant autumn was under way in New York City. The air, heavy with steam and auto exhaust, had cooled and was cut by the scent of fire smoke and roasted corn from the handcarts. The early light coming through the oaks turned the air gold. Flame-colored leaves littered the weed-choked grounds of Central Park, colorful fragments among the trash, dead trees, and underbrush that had overtaken the park in the last few years. Benches were turned over with tall wisps of grass growing between the slats. Large patches of dirt spotted the park where grass had died in the summer heat or where the sheep had overgrazed. Thousands of rats had colonized in the park. In spite of its condition, Central Park was one of the last natural refuges in a city blossoming with steel bridges, skyscrapers, trains, and subways.

Jelliffe was a thinking man, who liked long walks alone with his thoughts. On nice days like that autumn morning, he often walked from the Neurological Institute through the park to his townhouse and office. Entering the threshold of the gate was like literally stepping into the countryside. Jelliffe would have smelled sweet hay and horses as he crossed one of the many bridle paths, usually busy with the riders who kept their horses stalled in the park for convenience. He would have seen tufts of white sheep on the Green, which was now appropriately called Sheep Meadow. It was not just the rats, horses, and sheep that called the park home; the run-down menagerie on the east side of the park held all types of exotic animals, including those that traveled with a circus. Jelliffe would have passed the Casino, the elm-lined Mall, and the Dairy before stepping back out onto busy Sixth Avenue. There, he would enter the urban world once again with the el roaring overhead, a sea of black, glossy cars, and the smell of petrol, frankfurter stands, snuff shops, and fresh roasting coffee. On some days, Jelliffe ducked into one of the tearooms or cafeterias lining the streets during this Prohibition age, when so many restaurants had gone out of business. On other days, the pushcarts offered quick treats: soft drinks, Clark Bars, Baby Ruth candy bars, Yoo-hoos, Eskimo Pies, and Popsicles.

In spite of Prohibition, it was a short walk out of Central Park to the midtown speaks, which now greatly outnumbered the bars that had been scattered through the city prior to Prohibition. It is not known if Jelliffe himself visited speakeasies, but given the fact that he brewed his own beer and produced makeshift wine and scotch, he was clearly not one of those to strictly adhere to the Eighteenth Amendment. He was also on a panel of physicians who spoke at the New York Academy of Medicine in 1919 on the subject, and as the *New York Times* reported, "While admitting to the evils of alcohol, Dr. Jelliffe said that it had been of great social benefit, and that moderate drinking was particularly useful in counteracting the tendency of the struggle for existence to harden men. . . . " Jelliffe also noted the number of illnesses, or even deaths, occurring from poorly made bathtub gin and other homemade liquors.

New York's health commissioner, Royal Copeland, had even argued that Prohibition would increase an already growing drug problem: cocaine. Cocaine was still being used for medicinal purposes, such as treating melancholy, fatigue, or seasickness, so it was widely available. "In one month," Copeland complained, "one drug store sold 500 ounces of cocaine, enough to send 2,500 people to hell." What enraged him even more were those physicians who were bribed into writing one hundred to two hundred prescriptions of cocaine per day. Copeland opined that they should be "boiled in oil." Cocaine had even been a problem for New York during the war. Some eight thousand men in the city intentionally took cocaine as a way of dodging the draft and were rejected as drug addicts.

As Jelliffe crossed the city blocks from the park toward his townhouse on Fifty-sixth, the rattling of progress could be heard on every street as automobiles sputtered and sprayed mud or gravel. When the century began, automobiles were a novelty for the rich. By the 1920s, there were over 17 million of them in the United States. In a photograph from 1900, Fifth Avenue is seen covered in horse-drawn carriages, with only one automobile in the entire photo. By the 1920s, that scenario was reversed, like a photographic negative of progress itself. The smell of manure was replaced by the strong scent of petrol, and horsehair clinging to a woolen coat was a sign of the past. Concrete roads had been built to accommodate those autos, the vast majority of them black Model Ts. This endless procession of black cars was abruptly broken by splashes of yellow and checkered cabs. And the roads themselves had been lined with sidewalks to protect pedestrians and those modern tree trunks known as telegraph poles. As New York adapted to this influx of autos, the city landscape changed suddenly and dramatically. Cobblestones were paved over. Soon, nearly all boulevards would have their trees cut down and pavement laid to widen the roadways. Gas stations appeared on street corners. Sprawling houses were torn down and parking lots built.

Progress was not just evident above the ground, but beneath it. In the last twenty years, city designers had been building train tracks *below* ground. One forward-thinking New Yorker predicted, "surface travel will be an oddity in New York twenty years from now." Futuristic plans for the city also included second-story sidewalks made of glass, leaving the dusty and dirty streets to automobiles and double-decker buses.

For New Yorkers, for Americans, and for the world, the 1920s would prove to be the decade with the most rapid technological change in history. In one generation, travel by horse and carriage would make way for autos; people would travel underground, and soon, in the sky; wireless radio would change ship travel; kitchen appliances and indoor plumbing would become mainstream; light would come from a switch and heat through pipes; telephones would appear in the majority of homes; and the canned music and crackling voice of radio would provide home entertainment and news.

Of course, those modern conveniences would not reach all corners of society. A simple walk into the tenement housing showed exactly where the borders of modern life stopped. The faded, water-stained brick buildings had no indoor plumbing or electricity, a fact made more obvious by the clothing lines strung between the buildings, with linens, cloth diapers, and the limp legs of stockings waving like white flags. The monotonous façades of naked windows were broken only by the skeletal staircases fronting every floor. In more appreciated New York architecture, the façades of the buildings boasted ornate detailing, columns, eaves, and tresses. In tenement housing, it was the fire escapes.

With or without those modern conveniences, this modern life was creating a new kind of stress. The War to End All Wars ushered in America's most violent century in history, and Americans had just come out of a war that seemed to have no definitive finish, established no sound peace. The draft had also been enacted in 1917, changing attitudes toward war, especially one fought in another part of the world during a time of isolationism.

Those immediate dangers, however, were still across the Atlantic. It was on American soil that a new fear arose to add to the amalgam of modern stress, and it was known as the Red Scare. New Yorkers now passed armed police outside every church and federal or municipal building in the city. Anti-immigrant sentiment, particularly toward Germans, peaked during the war; but it was the Bolshevik revolution in Russia that was igniting fear in the United States. A flurry of legislative activity like the Espionage Act of 1917 and the Sedition Act of 1918 addressed these fears, but the laws only fueled more anarchist activity. In the spring of 1919, a plot to explode thirty-six bombs in the United States was thwarted just in time. On June 2, however, another anarchist plot was not uncovered fast enough, and eight bombs exploded just before midnight in eight different cities. The targets were judges, attorneys, and congressmen involved in anti-anarchist legislation. In New York, at the home of a judge, a bomb exploded prematurely on the front porch of a brownstone on East Sixty-first Street. The judge himself was not home, but his wife and the housekeeping staff were there when the explosion took out the front half of the home. A child came down the staircase just moments before it collapsed. Windows shattered, and iron was found embedded in the walls of neighboring homes. In spite of that, there were only two deaths, a man and a woman believed to be the bomb makers. The *Times* did not mince words with the morning coverage on June 3: "A man and a woman were blown to pieces this morning." The article went on to explain how investigators located body parts all over the block. Those explosions were followed a few months later by a large explosion outside of J. P. Morgan's bank on Wall Street.

In addition to the anarchist violence was the birth of bootlegging. The Five Points Gang, now boasting members like Al "Scarface"

Capone, Charles "Lucky" Luciano, and Frank Costello, was the most powerful in a number of crews that bloodied New York streets.

If violence was not enough to make New York feel unsafe, the latest progress in automobile mass production would. At a time when medicine was rapidly advancing life expectancy, one observer wrote, "The one appalling increase in the number of deaths, from a cause that is among the least excusable, has been in automobile accidents." Stepping off a curb could literally prove fatal.

Parents also worried about the children of this age. For the first time in history, children had a wealth of free time and free roam of the streets. They were not saddled with the hefty responsibility and chores of farm life. Modern food production, mass transportation, and appliances eased those burdens. Parents worried that their children were losing something valuable in the exchange. They stood by and shook their heads at a generation of children who spent their time on tops, marbles, and kick-the-can rather than learning riding, shooting, tool work, and other building blocks of self-reliance.

All of these changes, rapid and monumental, were affecting the American psyche. At the same time, the field of psychiatry was finally gaining the approval and acceptance of the medical community—in some sense *because* the stress of life had become so pervasive. Jelliffe was convinced that physical manifestations of these psychological worries plagued modern man. For neurologists in the 1920s, the focus was not so much on tackling every problem as it was on maintaining what they referred to as "mental hygiene." Just as personal hygiene helped control infectious disease, mental hygiene would keep the mind clean and healthy.

Surely that thought occurred to Jelliffe as his new patient, Adam, sat before him in the fall of 1924. Like the physicians in Europe—von Economo in Vienna and Hall in England—Jelliffe wondered about food poisoning when he encountered his first encephalitis lethargica patients. Jelliffe went so far as to suspect faultily packed olives. But, like the other physicians, he soon dismissed the idea as the disease took on new symptoms. By the time epidemic encephalitis had reached its peak in New York City, Jelliffe began to see patients who were not in the throes of the disease, but had sustained damage from it.

Adam had arrived at Jelliffe's office on West Fifty-sixth wearing a gray suit buttoned over a vest and a striped tie. A fedora was tilted on his head. Looking around the office, Adam would have seen the spines of dozens of medical journals, a desk cluttered with stacks of paper, and a patchwork of sunlight on the glass vials and jars. In front of the journals was one large bell jar and a microscope that gave the office an air of scientific research, as though peering through that great lens would shed any light on the darker confines of the mind.

The first thing Dr. Jelliffe noticed about the young man sitting before him was a well-marked and progressive type of Parkinson's disease. For the first three weeks Jelliffe saw him, Adam walked like a dummy, his arms drawn up to his sides, hands limp before him, and he showed a distinct tremor. Jelliffe diagnosed him as having the chronic effects of epidemic encephalitis.

Adam, Jelliffe recorded, was a young Jewish man about twenty years of age. His father was Russian, his mother Hungarian. Adam was one of five children and, according to his files, had been born without difficulty, walked and talked at the usual age, read by the age of five or six. He did not wet his bed, bite his nails, stammer or stutter, walk in his sleep, or show any other compulsive habits before his case of flu during Easter break of 1922.

Since that attack, he had suffered as many as three or four respiratory attacks a day. At times, lighting a cigarette would help divert his attention and stop an attack. At other times, lighting a cigarette could trigger one. During an attack, Jelliffe noted, Adam's eyes would dilate and his face would turn into a mask. Sometimes, his hands would clench and cramp like they were trying to grasp something. Adam described it as "Jesusly painful." He grew to dread the attacks so much that he could become anxious and cry just thinking about them.

Jelliffe started asking Adam to record his dreams. Ever the psychoanalyst, Jelliffe found Adam's dreams rife with sexual content, innuendo, and Oedipal tendencies. The fact that Adam had seen angels the night his delirium first started seemed clear from a psychoanalytical point of view: wooden angels were etched into the back of the sofa where his mother had nursed him as an infant. After particularly vivid dreams, including one in which a dog was biting and shaking his hand—an obvious symbol of masturbation in Jelliffe's analysis—Adam would feel rotten and spend the next day in bed.

Toward the end of the month, Adam arrived at Jelliffe's office after a hard trip from Philadelphia. He had gone into a trance in the taxicab over a fear of hitting an elevated train pillar. Adam came into the office bent over and stooped like an old man. Jelliffe patted him hard on the back and said, "Limber up! Brace up!"

Jelliffe noticed a look cross the boy's face like a shadow. He asked Adam what he felt just then.

"I wanted to say, Cut it out—God damn you! I hate you! You were so like father trying to make me get up in the morning. The goddamn son of a bitch!" Adam left the office agitated and angry.

Jelliffe was convinced there was a link between the stress still present in Adam's mind and the triggers for the physical reaction. Adam's particular case, it turned out, was not so unusual. Throughout Europe there had been strange respiratory tics occurring in epidemic proportions—hiccups, unceasing yawning, and breathing tics like Adam's. All of them had been triggered by a case of encephalitis lethargica. Jelliffe would later write, "No single situation in neurology has offered so much opportunity for the analysis of physio-pathological phenomena . . . as has epidemic encephalitis."

Epidemic encephalitis created an opportunity to study the effect of an organic disease of the brain damaging the mind—the very argument being made by neuropsychiatrists like Jelliffe. The disease injured the brain and caused physical repercussions that could not be controlled by patient or doctor. Yet the symptoms *could* change given certain personal circumstances. Patients responded to different types of stimulation and different types of people.

In New York in 1920, epidemic encephalitis kept a twenty-nine-year-old woman asleep for more than one hundred days. Knowing how she loved music, her husband hired a young violinist to play at her bedside. He started with one of Liszt's Hungarian Rhapsodies, to no effect; but when he played Schubert's "Serenade," the woman suddenly opened her eyes and remained awake. She had a full recovery. The *London Times* reported, "This is the first case in the records of the New York Health Department of a cure in a case of Encephalitis Lethargica."

Decades later, Dr. Oliver Sacks would encounter similar cases in his studies of encephalitis lethargica survivors at the Beth Abraham Hospital in the Bronx. Sacks was able to revive the patients briefly with the drug levodopa or L-dopa—a story he chronicled beautifully in the book *Awakenings*. Although Sacks encountered these catatonic, "extinct volcanoes" in the ward, he found they were anything but extinct. If thrown a ball, a patient would reach up and catch it. They would respond to certain pieces of music, but not others. If held by the hand, a patient could walk. Sacks found that the dysfunction was not in physical impairment or paralysis, but in the inability to initiate movement. The patient simply did not have the "will" to begin to move. Inside their brains, damage to the basal ganglia corrupted message signals like interference on a telephone line, and the automatic impulse to move was stunted. When Sacks or a member of his staff stood beside a patient and walked, the patient could follow, as though sharing the willpower to move. When let go, the patient would fall to the floor.

Sacks, like Jelliffe, also found that in many cases, the patient's own life and experiences played a large part in determining the course of this disease. As the brain is the body's most mysterious organ, and because no two people are exactly alike, no two cases of encephalitis lethargica were the same. This encephalitis was an epidemic disease—it was *spread* somehow. It was not inherited, and it was not the result of a traumatic personal event. Its damage to certain parts of the brain was physiological and visible in autopsy. And, yet, the disease could prey upon personal aspects of a patient's life. It blurred all the lines between an organic disease and the psychology of a patient. As Sacks himself said: "Psychiatry and neurology are inextricably linked. Defining it one way or the other doesn't do justice to the patient."

Sacks would also refer to Jelliffe as "perhaps the closest observer of the sleeping sickness and its sequelae."

Jelliffe found that certain authoritative personalities triggered respiratory attacks in Adam—usually having to do with Adam's strict father. Once Jelliffe knew this, he was able to work to great effect, analyzing Adam's responses and helping him understand his emotional and physical reactions. For Adam, just understanding what prompted these tics and symptoms must have given him some comfort. After years of feeling out of control of his own body and mind, Adam could at least understand what was happening and why.

In 1925, Jelliffe wrote a letter to "My Dear Dr. Freud" and described Adam's case and progress; he even described Adam as "practically completely restored." Jelliffe went on to say, "He was a severe Parkinsonian with respiratory attacks and all the character changes that threatened a deteriorating psychosis, and I only analyzed his efforts to use regressive mechanisms in your sense . . . and have had a brilliant therapeutic result."

If Jelliffe had any failings in his study of encephalitis lethargica, it was his optimism. The mind healed, the brain might still be broken.

CHAPTER 11

Only the Beginning

In June 1927, an energy coursed through America, electric and hopeful, an upwelling of pride and awe. On June 13 in New York City, that energy literally filled the air like a blizzard. With the world watching, Charles Lindbergh had just completed his solo trek across the Atlantic from New York to Paris. When he returned, thousands of people lined the sidewalks while stock exchange ticker tapes snaked through the air and the gray confetti of shredded newspapers rained on their upturned faces. People stared skyward at the shredded cotton clouds and tried to imagine how a man was able to fly such a distance. Through the crowds, the gaunt figure of Lindbergh was barely visible through the snowstorm of confetti as he sat on the back of the auto trudging through the crowds. The celebratory spirit continued all week, and Adam, like most other young people, probably took to the rowdy streets. There were parties throughout the city and people had come to New York just to see the return of Lindbergh. Brass bands played on sidewalks, and traffic police wearing white cotton gloves directed crowds. In a city that already echoed with traffic whistles, ship horns, shrieking el trains and streetcars, car horns, and church bells, the sounds seemed even brighter. As the parade ended, a whole fleet of street cleaners wearing their white wing uniforms and carrying sweepers hit the streets.

That same week, run-down and tired, Adam caught a simple fever, a "febrile attack" as it was noted in his case history. He had been out the night before, celebrating with friends and, as his psychoanalyst Jelliffe was quick to point out, had an "amatory adventure" with a young lady. He woke up the next day "shaking like a leaf."

It had been five years since Adam first saw Jelliffe. The psychoanalysis and psychotherapy had been successful, and Jelliffe included this case among others in an abstract he read at the Fifty-fourth Annual Meeting of the American Neurological Association in Washington, D.C. It was early May when he read his abstract, "Psychologic Components in Postencephalitic Oculogyric Crisis," and then the paper was presented for publication in June. As the final proofs were going to print, Jelliffe was forced to make a sudden and significant change to Adam's case history.

Other physicians had warned of the tendency of encephalitis lethargica to have long-term side effects. Someone could suffer an acute case of epidemic encephalitis—often immediately following a case of the flu. Or sometimes the encephalitis followed a sore throat and a fever. The deep sleep usually came next. And that was the end of the acute part of the disease. In Adam's case, like others, he awoke from that acute phase a different person—something he and his doctors struggled with for years, often including compulsive behavior, tics, stiff movements, or fits of delusion.

As the 1920s progressed, with the epidemic still running its course through the world, the first of those patients began showing what would be the hallmark of this disease: chronic symptoms.

It could be weeks, months, or years after an attack of epidemic encephalitis that new, frighteningly similar symptoms surfaced. It became one of the strangest and most terrifying aspects of the disease, leaving thousands, if not millions, of victims institutionalized for the rest of their lives.

Adam had been fine, fully recovered, from his case of epidemic encephalitis and the symptoms that had immediately followed. He became normal once again—sleeping well at night and waking at around ten or so in the morning. Then he would go to his father's store, where he worked as a junior salesman, and stay until midafternoon. He would take an afternoon nap before returning to work in the evening or go out at night with friends. His family noticed that he seemed to have "grown up" over the last couple of years, though he was still in his midtwenties.

In June, when Adam's symptoms returned, his brother called Dr. Jelliffe in a state of panic. Adam had gone into an "oculogyric crisis," one of the most common chronic symptoms of epidemic encephalitis. His eyes locked upward and sideways in their sockets, his neck and head stiffened, his shoulders hunched upward, and he started going into spasms. To someone watching, it has the terrifying quality of a seizure—a sense that the human body is very much out of our control, and the person inside seems momentarily lost or disconnected. Once again, Adam described it as "Jesusly painful." It lasted for the next ten hours.

According to Adam's brother, the attacks came on when he was exhausted or under duress. He had been free of problems for so long, the onset of this disturbing new symptom was very stressful to Adam, worse somehow than the breathing tics and rigid movements and trances he had once been prey to.

"I guess I just deserve this," Adam would say. "Everybody gets what they deserve."

The seizures became so distressing to Adam that the worry alone could bring one on. He returned one evening from his father's store and an oculogyric crisis ensued. His brother asked, "What started your eyes tonight, Adam?"

"I keep thinking about it and worrying and they go south," he replied.

His brother recorded all of these instances in a letter and sent it to Dr. Jelliffe. Near the end of the letter he wrote, "He certainly is a game little kid and it almost breaks my heart to see him have this new trouble. I hope you will find a way to handle it."

Adam and his brother arrived at Dr. Jelliffe's office to discuss what to do next. The walk was an easy distance from the Sixth Avenue el, or down Fifth Avenue, although the street was growing more crowded now that Bergdorf Goodman was under construction. For the time, it left the plaza with a huge open cavity where Cornelius Vanderbilt II's chateau mansion once stood. Once onto Fifty-sixth, it was a quieter walk along the row of townhouses to Jelliffe's home and office. Inside, Adam and his brother passed an older woman, crouched over a typewriter, who had been a former patient of Jelliffe's and, as a true believer, had gone to work for him when she recovered. Adam again sat in the office with the blanched white tablecloth, bell jars, a microscope, spines of journals. Staring at the long rows of medical journals, it must have seemed sad, even desperate, that there were no answers in them for Adam. Outside, cars honked and pushed slowly through the crowds of pedestrians. Hammering and saws could be heard from the construction outside, and flatbed trucks with rope-tied construction supplies rattled up the street. The house shook each time the el passed.

Jelliffe was disappointed by Adam's latest attack. He looked at the young man, asked him questions, and talked about options. It was late in the afternoon; twilight was crawling across the office, and the light flushed with dusk. As Jelliffe asked him standard questions, Adam went into another trance and seizure right there in front of the doctor. Had it not been dark, Jelliffe would have photographed the episode for his study, but with the light too dim and Adam's face eclipsed by shadow, Jelliffe could only watch. Adam's eyes rolled upward into his head, his body froze, his neck stiffened, his head dropped back, and he began murmuring from the corner of his mouth, "A million ideas are going through my head, rape my sister, rape my mother, kill my brother, kill my father." "Am I going crazy?" he whispered.

Jelliffe could only helplessly watch the episode that possessed Adam so completely. It was the thought of going crazy that brought on the attack, the uncontrollable, insane, violent thoughts flashing through his mind. They left Dr. Jelliffe's office, and his brother took him home. People on the streets stopped and stared as they hurried by. Adam's seizure lasted another thirty-six hours.

Jelliffe wrote in his report that "this is a typical example of what occurs in other patients, with many variations." At times, it even leads to suicidal thoughts, and it wasn't long before Adam made an unsuccessful dash for the window in Dr. Jelliffe's office to try to end all the thoughts once and for all.

Over the next several months, Dr. Jelliffe saw Adam less and less. He was never sure if that was due to lack of money or from the stress brought on by the visits. Jelliffe believed that Adam had a "positive transference" with him, but that too many factors in his life interfered. There is no way of knowing definitively whether Jelliffe's psychotherapy helped Adam's case, hurt it, or had any effect at all. Nor is it known what happened to Adam—whether he recovered enough to be able to live a somewhat normal life, or if he, like so many others, ended up in an institution.

Adam's case was hardly unique. Another neurologist in New York wrote in one journal, "That mysterious disease, the real cause of which is still unknown, which remains dormant for many years, recurs at long but irregular intervals, and leaves after each outbreak a trail of broken minds and crippled bodies." Neurologists throughout New York, the United States, and abroad were beginning to publish articles about similar cases. By the end of the decade, nine thousand articles on encephalitis lethargica appeared in the medical literature.

Epidemic encephalitis was approaching its peak in the United States. In Europe, it had been a steady build, a slow burn. For the most part, the cases in Europe had been sleep cases or insomnia cases, but very little was known about the aftereffects of the disease. It was still too new. In Vienna, the birthplace of the epidemic, cases had disappeared. In the United Kingdom, however, cases were becoming more alarming. Not only was the mortality rate soaring toward 56 percent, but the survivors awoke as new, terrible versions of their former selves. Doctors began including in their published studies reports that the disease was said to "alter the dispositions of those who have recovered from it." It was said almost hesitantly; they were not sure what it meant. Or they did not want the public to know how much worse it could get.

By the time the epidemic peaked in New York, the public, the press, and the health department were still chasing the shadows of this disease. The health department could reveal little about how the disease was spread or how to protect oneself from it. There were no notices to tack up in neighborhoods or educational meetings they could hold for the public. They didn't have enough answers themselves. Even tracking its spread was proving inconclusive. They tried in vain to find a pattern, to trace this epidemic the way they had polio in 1916 or influenza in 1918. But epidemic encephalitis followed no particular time frame, and unlike diseases that target the impoverished, immigrants, or soldiers, this disease showed no discernible class preference. It rarely even spread within one household. For medical investigators, it was like trying to catch the pattern of a snowflake before it melts in an open palm.

The newspapers made only occasional mention of the disease, and even then it was usually to report some oddity or declare the longest-running sleep so far. The newspapers were preoccupied with the idiosyncrasies of the disease and, as a result, were grossly underestimating it. Epidemic encephalitis was about to attack one of New York's most prominent families, securing its place in the pages of the *New York Times* and bolstering the work of neurologists.

CASE HISTORY FOUR

East Island, New York, 1925

NAME: Jessie

PHYSICIAN: Dr. Frederick Tilney

CHAPTER 12

Jessie

The yellow summer heat had settled over the city, bringing higher temperatures during the day and nights that were still cool and breeze-filled. While the weather was still mild enough, handcarts sold fresh lemonade and malted milk. Soon, however, those cooler temperatures would give way, leaving the homes and buildings of New York with hot, weary air, and everything about the city would seem to move at a slower pace, people trudging through the heavy heat of summer. The brackish waters of the Hudson gave off a salty air and feeling of imminent escape. And on weekends the empty city became, as F. Scott Fitzgerald described it, like overripened fruit.

In June 1925, Dr. Tilney made his way through the sprawling, two-block, colossal Penn Station. Though the station boasted majestic stone columns and a long lip of shallow steps, the concrete and stone façade betrayed the interior. Inside the light-filled train concourse were vaulted ceilings of glass and a network of arches, a throwback to the Crystal Palaces popular in the previous century. Newsstands carried newspapers and magazines with wholesome images fanned out like the feathers of a peacock. *Ladies' Home Journal* showed a color illustration of a bride and groom; *Good Housekeeping's* cover had a shot of a mother reading to her daughter; *Field and Stream* showed an image of a man and woman having a picnic beside a stream; the *Saturday Evening Post* presented a Norman Rockwell sketch of a woman lounging on a chair; there was even a new weekly called the *New Yorker* on the stand.

As Tilney waited to board the train, sunlight flooded the station, and the iron latticework left a web of shadows across the floors. The sky was cloudless and clear, the color of pale blue porcelain. At any other time, a trip to the Long Island shore on a beautiful summer day would have been a pleasant occasion. But that was not the case on this particular afternoon.

It was a Sunday, and most weekenders were returning to the city after a respite from the heat along the coast or in the country, so the train out to Long Island was almost empty. At times, the commute back into the city was so clogged with automobiles, and the infrastructure of roads so ill-equipped for traffic, that people left their cars on the side of the road and took the train or subway home. The train, too, was hot, but when the windows were lowered, the cooler air carried in the smell of sweet grass and clover. Wild carrot bloomed along the tracks and among the tree trunks.

A driver picked Tilney up at the station for the drive toward East Island. Tree-lined roads offered shade and scattered, kinetic light that skittered along the pavement. The auto slowed and pulled onto a long cement drive over a bridge guarded by a high wrought-iron gate and two wardens. At the end of the long drive, lined with linden trees, a great Georgian estate loomed in the distance. It was three stories tall and made of brick, with a porch on the lower level. Throughout the grounds were well-tended gardens and musky roses in bloom; the estate had several dozen gardeners. Even the grand stairway in the entry held fresh flower arrangements.

Tilney hurried into the house and was shown to a second-floor bedroom where a local physician, Dr. Everett Jessup, waited for him. Jessup had already made a diagnosis of sleeping sickness, but given the circumstances, he wanted a more experienced opinion as well. Epidemic encephalitis had come suddenly into Long Island, killing six people the week before, and five people the week before that. The disease had also made world news the month before when Viscount Milner, Britain's former Secretary of War, died of encephalitis lethargica.

Tilney reviewed the notes on the patient's charts, and he looked at the woman now languishing in the bed before him. Tilney knew that this was a case that would make international news. He wanted to give a sure diagnosis before the staff issued a radio message to the yacht drifting in Long Island Sound, telling J. P. Morgan his wife had sleeping sickness.

Mrs. Jane Morgan, called Jessie, had gone to St. John's Church of Lattingtown on that June morning. It must have been an unusually beautiful summer day because her husband, Jack, spent most Sundays at the church, usually passing the collection plate. On that Sunday, however, he had taken his yacht, the *Corsair*, out into the sound and planned to pick Jessie up that afternoon. But, during church, Jessie had turned to friends in the pew and said she felt ill, as though she might faint. A few minutes later, she mentioned again feeling faint and left during the service. She found her auto, and her driver hurried through the several miles of road that lay between the church and the estate on East Island. As they snaked their way toward the house and pulled into the long, linear driveway, her head began to ache violently.

Jessie was carried into the house, taken to her second-story room, just above the library and ground-floor porch, and put to bed. Her condition declined rapidly and her personal physicians were called.

A summons was flashed across the wireless telling Morgan to return home immediately. He arrived by sundown and went straight to her bed, where Jessie was barely conscious and severely ill. There could not have been a greater blow to Jack P. Morgan.

Jessie and Jack Morgan had been married for thirty-five years and had two sons and two daughters, all grown. Both Jessie and Jack were loving parents, but they were devoted to each other even more. Their children would later say that the couple was so close, the children themselves felt excluded at times. Although their marriage had bound two prominent families together, and their wedding made front-page news in the *New York Times* and the *New York Post*, theirs seemed to be a genuine love affair.

Jack had originally wanted to be a physician, but was pressured into the family banking business. Perhaps, then, his marriage and family were the only real source of happiness to him. It is also likely that Morgan had grown up watching his philandering father wreck their family life. Whatever the reason, Jack Morgan remained a faithful and steadfast husband. In his biography of the Morgan family, Ron Chernow described their marriage: "She propped up his ego, and he relied implicitly on her judgment in many matters. Jessie was strict with the four children and ran the estates with a firm, expert hand. She was cool and businesslike. . . . But to Jack, Jessie was the supportive presence who compensated for his lifelong insecurity and guaranteed he would be spared his father's terribly loveless fate."

Jessie was known to be an independent, confident, and decisive person. She was described as feminine, but strong to the core. Still, Jessie Morgan kept completely out of society life. Her generous philanthropic contributions were made anonymously. And her life revolved entirely around her husband, her children, and her gardens—all three of which she tended with expert care.

One morning in 1915, a crazed gunman made his way past the butler into the entry of the Morgans' home. When the Morgans heard the commotion and came down the stairs, they found the gunman, holding their daughters, pointing two pistols up the staircase. Without hesitation, Jessie Morgan threw herself at the gunman and attacked him. Jack Morgan managed to push his wife out of the path of the first bullet, but was hit himself. When the gunman struggled with Jack over the gun, Jessie again stepped in, pinning down his arm and preventing him from firing a second shot. After the gunman had been subdued, they found a large stick of dynamite in his pocket as well. Jack always credited his wife with saving his life.

Now, Jack Morgan stood at his wife's bedside as she struggled for her own life. As the light grew dim and the sky copper-colored, night approached, and Jessie Morgan fell into a deep sleep. It was decided then to give her a blood transfusion. A message was sent to St. Luke's Hospital in Manhattan to find a donor, and the next morning a surgeon named Dr. Bishop arrived to donate eight ounces of his own blood. The operation took place at ten o'clock that morning, while Jack waited in the next room. The transfusion was successful, and because of the swelling in the brain, intravenous dextrose was given to Jessie Morgan as well. The effect was immediate improvement. Every morning after that, for the next nine weeks, Jessie would receive the dextrose treatments. As Jessie slept indefinitely, she also had to be fed by a tube three times a day with pureed vegetables or creamed soups. And, not taking any chances, Jack Morgan assigned five physicians and five nurses to her care. With alternating schedules, someone was with her at all times.

Tilney routinely made trips to Glen Cove, not only as an expert on epidemic encephalitis, but as a trusted family doctor—he had been the personal physician of Jack's mother, who had died the year before. Tilney and Jessie's other doctors believed that she might have contracted the illness through the close contact she had had with several grandchildren who had the flu that spring.

Jessie's health seemed to improve, and Jack returned to the bank at 23 Wall Street, tacking up bulletins every day on his wife's condition. He wrote to a friend: "Jessie is getting on well, the doctors assure me, and they all assure me that the recovery, though very slow, will ultimately be complete. . . . Of course no one can tell how long she must sleep, but while she sleeps she is not conscious of any pain or discomfort, and the cure is proceeding all the time."

With Jessie improving and a staff to watch over her, Jack left for Manhattan on the morning of August 14. Late that morning, he received a phone call from home telling him that Jessie had taken a turn for the worse. He raced out of the bank to the docks and onto his boat. As the boat approached Long Island's Gold Coast, the vaulted rooftops, chimneys, marble archways, and blocks of silvered window-light rose above treetops and boxed hedges. The coastline was so glamorous that it had been the inspiration for a novel just released that spring, *The Great Gatsby*. To anyone else, Jack Morgan, one of the wealthiest men in America, drifting on a boat toward an estate on his own island would seem to be a man who had luck on his side.

As the boat dropped speed and edged toward the dock, Jack could see the servants spread out along the coast to keep reporters or curious bystanders from sneaking onto the property. The only other way onto the estate was through the heavily guarded, gated bridge, which had been under the watch of a superintendent ever since the 1915 break-in by the crazed gunman. Jack climbed out of the boat and hurried across the lawn, only to be met with the news that Jessie had already died, at the stroke of noon. Her heart had stopped.

Another radio summons went out to the *Corsair*, but this time it was for Jessie's two sons. As they made their way through Long Island Sound toward home, a storm was gathering to the east. A deep, gray-blue ceiling of clouds narrowed the distance between the water and the sky. Wind whipped the tree limbs, and lightning lit the underbelly of clouds every few minutes. By the time her sons arrived at Glen Cove the severe storm had knocked out power at the island's electrical plant. Their home was alight with candles in every room, like a house burning from within.

After her death, a notice was posted.

> Mrs. Morgan, who for the past two months suffered from lethargic encephalitis and had slowly improved until a few days ago, died at noon today as the result of a sudden cardiac arrest.

It was signed by her doctors, including Tilney.

The funeral for Jessie Morgan was private and for family members only, many of whom arrived directly from their summer homes on that hot August afternoon. The services were held at the ivy-covered Gothic church, St. John's, where Jessie had first become ill. The mourners filed in wearing black crepe satin or silk, veils, gloves, and fans for the heat. She was buried in a family plot beneath white oaks and maple trees. Western Union had to send an extra telegraph officer to the Glen Cove office to manage the flood of condolences. Guards were employed to stand along the roads and hold back crowds as the Morgan family drove past in a procession of polished black limousines.

Epidemic encephalitis, as it turned out, formed a sad coincidence between two warring families. Both Jack's father, Pierpont Morgan, and his grandfather, Junius, had fought bitterly with the Rothschild and Sons banking empire in London. It was a multigenerational conflict. The sons and heirs to their fortunes, however, had a few things in common. Both Jack Morgan and Charles Rothschild were sons dutifully sent into the family banking business, in spite of other loves—Jack's was medicine and science; Charles's was entomology, and he compiled an incredible collection still carried in museums today. Both sons seemed to lack the vigor and ruthless ambition of their fathers. And both men would find themselves horribly touched by sleeping sickness. Jack lost his beloved wife in 1925, and Rothschild developed sleeping sickness after surviving a case of flu during the influenza pandemic. To his wife and children, Rothschild seemed depressed and frustrated by the encephalitis, but not a danger. It was a terrible surprise then when Charles Rothschild locked himself in the bathroom of their home in Ashton Wold and cut his own throat. It was ruled a "suicide during temporary insanity."

Jack Morgan grew depressed after the loss of his wife. He spent several million dollars to buy up adjoining properties along the Gold Coast, as well as a boathouse, to build a memorial park to his wife. Friends later remarked, after visiting East Island, that the house had the eerie sense that Jessie was still there—Morgan was refusing to let her go. He kept her bedroom exactly the same and tended to her garden. Just that spring, Jessie's tulips had won first prize in a local contest. Morgan wanted to be sure her garden would come up the following spring, in spite of the fact that she would not be there to see it.

It would have taken a brave person, or at the very least an ambitious one, to approach Jack Morgan in that state. But Tilney did. He asked if Morgan might make a donation to the Neurological Institute in his wife's name in order to support research for this disease. Morgan donated $200,000 to the institute, practically funding an entire floor of research for the mysterious disease that had taken the life of his wife.

CHAPTER 13

1925

In most cases, history can hold up a magnifying glass to an epidemic and look at the medicine, society, politics, philosophies, and religion of the time period. Encephalitis lethargica was just the reverse. The 1920s and 1930s became the great lens through which it was possible to see a decade-long epidemic, the victims, the survivors, the medical investigators who fought it, and ultimately, its vanishing from memory.

To understand how a disease like epidemic encephalitis could gain so much attention from neuropsychiatry, medical research, and the city health department so quickly and so efficiently, it is important to understand the era in which it happened. It is equally important to understand how this disease escaped the notice of almost everyone else. Epidemic encephalitis managed to fade behind the brightness of one decade and the darkness of the next. Unlike other major epidemics, this one was diffuse, borderless, hard to trace, impossible to define.

* * *

Most historians now consider the 1920s an extension of the Progressive Era. The movement began with the antitrust laws that promised to rein in some of the business megamonopolies, but it soon grew into the progress laws and ideals that aimed for a more modern, safer, and healthier lifestyle. Legislation like mandatory meat and milk inspections, restrictions on tenement housing, improved working conditions, and child labor laws passed. The movement then morphed into legislation concerning civil responsibility, granting Native Americans citizenship, and allowing women to vote, as well as the amendment disallowing alcohol; in an upwelling of civil rights concerns, the right to vote was given, and the right to drink was taken away.

Prohibition also ushered in the age of the bootlegger and the gangster, particularly in cities like Chicago and New York. Incidentally, those were also two of the cities hardest hit by epidemic encephalitis. One newspaper editorial put forth the theory that the rise in lawlessness and gangsterism was, in part, due to the moral degeneracy caused by epidemic encephalitis. A British professor of psychological medicine later asserted the same theory about gangsters, John Dillinger in particular, though there was never proof Dillinger had the disease.

As the progress laws gathered momentum, dramatic change also struck the White House when, in 1923, President Warren G. Harding died suddenly after only two years in office. Doctors never performed an autopsy, and it was never determined whether he had died of a heart attack or stroke. The mystery surrounding his death led to conspiracy theories about poison that implicated everyone from political enemies to his own wife, who knew of his long-term love affair with another woman. The Teapot Dome scandal was the Watergate of that time period, and further scandals arose following Harding's sudden death. Vice President Calvin Coolidge was visiting his family home in Vermont when a messenger arrived with the telegram—the home had neither a telephone nor electricity, so the thirtieth president was given the oath of office under kerosene lamplight by his father, a notary public. That act alone seems symbolic of the momentous changes taking place in America. Coolidge was sworn in with neither a phone nor electricity at hand; but his successor, Herbert Hoover, would take his oath of office with aircraft flying overhead to mark the occasion.

While the progress laws seemed a giant political step forward, a darker side to the progressive change arose as well, primarily targeting immigration. As rural areas of America moved closer to xenophobia, and the Ku Klux Klan spread throughout the country with millions of members, cities like New York did not escape the anti-immigrant sentiment. The National Origins Quota Act, and then the Johnson-Reed Act, intended to overturn America's open-door policy and return the country to its original religious and ethnic mix before the great immigration boom. There were more Italians in New York than in Rome, more Germans than in Berlin, and more Irish than in Dublin. A great influx of Puerto Ricans had also arrived following the Spanish-American War, and there was a massive migration of American blacks from the South to New York City, settling primarily in Harlem.

Some historians even consider progressive laws like women's suffrage and Prohibition to be as much about immigration as they were about civic well-being. Even birth control became a conservative cause, with education at its mildest end of the spectrum and the practice of eugenics at its most extreme. Suffrage was supported by educated, religious-minded women who hoped to have a majority influence on politics; their immigrant counterparts were not as likely to follow politics or vote. Likewise, birth control was in one sense a way to stem the growth of immigrant families, who typically had more children. In other words, white Protestant women could add votes to those of the white Protestant men. Prohibition, or the temperance movement, also took direct aim at immigrant cultures by targeting the German beer gardens and Italian and Irish pubs.

Even religion was playing a part. There was a major push to "Americanize" immigrants, particularly in northern cities like New York. Catholicism and Lutheranism were the religions of immigrant cultures and attempts were made to close parochial schools.

The immigration laws had a definite cultural effect on New York, but it was not the one originally intended. The laws changed the cultural topography of the city, but rather than becoming "Americanized," the immigrants blended into and ultimately enhanced New York. With second- and third-generation immigrant families easing into American culture, immigrant enclaves began to disappear. The city became less a series of segmented neighborhoods, each dominated by one ethnicity, and more of a whole, integrated city. Rather than living in neighborhoods that revolved around one church, one school, and a handful of vendors, existing almost like small villages within a larger framework, people began to spread out. Mass transportation, primarily the subway system, was facilitating that change. Soon, the diversity beginning to form was of a different nature. Writers flocked to New York, and the city published more books than Chicago, Philadelphia, and Boston combined; there was the Harlem Renaissance, especially in music; theater continued its rise; fashion and department stores flourished. New York was becoming the "City of Cities."

Part of what was making this change possible was the evolution in public health. With greater public health services and more knowledge about how disease spread, immigrants and their neighborhoods ceased being the scapegoat for the catchall nexus of epidemics, so there was less resistance to integrating neighborhoods and schools.

★ ★ ★

Against this rapid change in politics, technology, and culture, an almost unnoticed epidemic was proceeding at a sluggish pace. An outbreak of influenza, cholera, smallpox, or polio spread with impressive speed. This one, on the other hand, was too easy to overlook, too easy to forget, and that's what made it all the more dangerous. As doctors and public health officials focused their attention on acute cases of sleeping sickness, a more grotesque side of the disease was taking shape.

In New York, the rising number of cases of epidemic encephalitis and the notoriety of some of its victims gave the Neurological Institute a steady influx of patients. The American Neurological Association had also celebrated its semicentennial in 1924. It had seen fifty years of rapid change, from phrenology and rudimentary measurements of the skull to the splintering professions of psychiatry, neurology, and neurosurgery—each finally coming into its own by the 1930s. The association wanted to capture that monumental change and the history of neurology in America in a book, and it asked only two neurologists to compile the volume: Dr. Frederick Tilney and Dr. Smith Ely Jelliffe.

Tilney was a natural choice for coediting a book on the history of neurology, but choosing Jelliffe showed the shifting attitude in neurology during that time. Although Jelliffe founded and served as editor of the *Journal of Nervous and Mental Disease*, and continued in that role for over forty years, he also had fallen out of favor with most neurologists, even being asked to step away from his involvement with the Neurological Institute. His reputation as a psychoanalyst had grown and was well established. Jelliffe's energetic, almost manic devotion to psychoanalysis and psychiatry at first unnerved medical professionals. In time, however, his opinions were becoming a valuable addition to brain study.

In the midst of seeing patients, frequenting the Neurological Institute, publishing articles, and compiling this book, Tilney and Jelliffe, both advancing in years, were working long hours. It was taking its toll. Jelliffe lived with his second wife in his home on Fifty-sixth Street, while Tilney and his wife, Camille, lived around the park, on Fifth Avenue. Their children were grown and had moved on to their own families.

On a personal level, the opening weeks of 1925 would bring a tragedy for one of the physicians, and the closing weeks a tragedy for the other. On January 22, 1925, the second great heartbreak of his life struck Jelliffe. The first had been his wife's death years before. The second was the death of his son, William Leeming. Leeming, something of a golden boy, was a medical student who had recently graduated from Yale and married just three months before. The younger Jelliffe had been a swimmer for Yale and was known throughout the country, even trying out as an Olympic hopeful. On that night, Leeming Jelliffe returned from the Metropolitan Swimming Championship at the New York Athletic Club. He had the habit of tossing small objects back and forth in his hands when he was happy, and that night at the Athletic Club, he'd been tossing a water polo ball as he closed his locker and left.

When he arrived home, he said hello to his new wife in the kitchen and went upstairs to change into his dressing gown. It was then that his wife heard the shot. She ran up the stairs and found her husband, shot in the head, with a pistol on the floor.

The police and an ambulance were called, but they stalled in the snow. The policemen and physicians ran the last several blocks on foot and arrived at the home just as Jelliffe did. He had his son taken to Roosevelt Hospital and specialists called in. They performed surgery, but his son died just after midnight from a wound to his head—the very part of the body Jelliffe had spent his career studying. As Leeming had been happy, accomplished, and recently married, suicide was ruled out. Friends believed he most likely shot himself by accident—tossing and twirling the pistol, which he kept in his clothes bureau.

At the end of that year, Jelliffe would write to Freud about his son. Freud, after all, could sympathize; he had lost his beloved daughter Sophie to pandemic flu five years before. "He was 25, fatherhood apart, quite a remarkable boy," Jelliffe wrote. "He always did things thoroughly from childhood—had distinct mechanical gifts. He was in his second year in medicine. Lovable, sociable, helpful, and as sound a student as he was an all-round athlete. This thing almost finished me. . . . I know I am coming round—as my dreams of his being with me are becoming fewer."

Tilney, like Jelliffe, had been working at a frenzied pace—with his own patients, those in New York as well as those outside the city, and on his research, teaching, and fund-raising for the Neurological Institute. He worked fifteen hours a day without stop, often skipping meals.

Tilney's frantic work schedule did not seem to diminish his social life. As a testament to his high standing in New York, Tilney was among New York's elite in the boxes of the Metropolitan Opera House to see the famous Eleanora Duse perform *The Lady from the Sea*. It was her first performance in the United States in twenty years. The *New York Times* listed the impressive attendees at the opening night performance: Rockefellers, Astors, Morgans, Macys, Marshall Field III, Rose Kennedy, Theodore Roosevelt—and Frederick Tilney.

With his life moving at an unmatchable pace, his body could not keep up. It had been four months since his patient Jessie Morgan died, and a few days before Christmas, Tilney suffered a cerebral thrombosis, a blood clot in the brain. He recovered, partially, but was unable to speak for six weeks. In a few months' time, Tilney taught himself how to scribble using just his left hand, and, ever the physician, he kept detailed notes on the course of his own recovery.

Time magazine carried the remarkable story of what happened next. Tilney had a devoted following at the Neurological Institute, and doctors from the institute took turns sleeping at Tilney's house every night. Five of them at a time would carry Tilney's heavy, inert body back and forth from his bedroom to his study so he could work.

Scrawling only a few words per page, it took him merely months to write his two-volume masterpiece, totaling 1,100 pages, *The Brain: From Ape to Man*. The book, published in 1928, was called "the finest piece of evolutionary writing since Darwin." Almost as impressive, given the outcome of the recent Scopes Monkey Trial, was the fact that Tilney managed to publish on the subject of evolution without polarizing science and religion. Several hundred Protestant ministers endorsed his book. Its popularity also shows, once again, how differently the brain was perceived at the time. Tilney was able to study the brain not only as the most important part in the machinery of human evolution, but also maintain the mind's role in well-being, even calling it the organ responsible for a man's salvation.

Still, it must have been hard for Tilney to describe and write about the evolution of the brain, about the development over time of walking, talking, and especially the use of hands. "With the brain to direct its action," wrote Tilney, " . . . the hand became the master key opening all of the ways leading through the new and vast domain of human behavior." And, yet, Tilney had lost his natural ability for all of those things, having to learn them again even as he committed their importance to paper.

In spite of their personal setbacks, both Jelliffe and Tilney were caught in the midst of revolutionary changes in brain study just as the epidemic of sleeping sickness peaked in New York City. From the perspective of 1925, with so much momentum, energy, and change taking place not only in neurology and medicine, but in the world, no one could have foreseen the clouds beginning to gather.

CHAPTER 14

A Two-Headed Beast

When the epidemic of encephalitis lethargica first swept across Europe, physicians knew little of what to expect from this disease with its kaleidoscope of symptoms. The only thing it seemed the doctors could count on was that this disease was unpredictable. Jelliffe himself had said, "There is probably no other acute infectious disease which gives rise to, or results in so many diversified types of mental disturbance."

After the initial wave of acute cases that spread to countries all over the world from 1916 well into the 1920s, there was a steady decline in the number of acute sleeping sickness cases. By the late 1920s, the epidemic was considered if not over, at least on the way out. Physicians, both then and now, estimate that one-third of the patients died during the acute stage, one-third recovered completely, and one-third developed permanent disabilities.

Unlike polio, which had left thousands of children paralyzed, epidemic encephalitis often had no direct, comprehensible link to the illness that had prompted it. Children and adults paralyzed by polio recovered from their fever with immediate knowledge of what they had lost. Epidemic encephalitis seemed almost crueler in its legacy; it allowed patients and their families to hope there had been a full recovery. It was then that the disease's long-term effects came on suddenly and horribly. For the next several decades, these survivors of the epidemic, mentally or physically handicapped by the disease, would fall under the antiseptic term "*postencephalitic.*"

In many cases, usually the adults, it began with a tremor, a small, seemingly inconsequential movement; its implication, however, was anything but inconsequential. It meant that inside the brain, the wiring was beginning to fail, like small fuses beginning to blow. The neurotransmitters failed to deliver the right messages, and in most cases, those neurotransmitters were the ones that controlled movement.

In the coming decades, these patients developed an extreme form of Parkinson's disease, different from Parkinsonism as it is known today. Those slight tremors progressed quickly, and the uncontrollable movements soon tensed to the point of stillness, with the brain unable to tell the body to move. It came on slowly, as though the patient felt each of his or her limbs freeze as solidly as tree branches encased in ice. The disease left patients, for all intents and purposes, catatonic—completely unable to move, communicate, or care for themselves. During the encephalitis lethargica epidemic and in the decades immediately following it, two-thirds of all Parkinson's patients were postencephalitic. Even more striking, the average age for onset of Parkinson's disease during that time was thirty-six years old.

Parkinsonism became the most common symptom among the adults who survived epidemic encephalitis, but children experienced chronic symptoms of their own. It is not clear why the disease was having such a different effect on children than on adults. Some doctors believed adult brains were developed enough to maintain self-control. The brains of children, on the other hand, were not finished growing. In a way, their brains were still pliable. When the disease caused swelling in the brain of a child, it damaged the wiring leading to the frontal lobe—the part of the brain that controls personality, impulse, self-restraint, pleasure, addiction, and decision making. The frontal lobe does not stop growing until around age twenty, so severe or long-term swelling caused by something like encephalitis lethargica would stunt the growth in many of those behavioral areas.

Seventy percent of children who survived epidemic encephalitis showed psychological changes. A third of those survivors showed severe or dangerous behavioral changes. This disease was a two-headed beast. The fatal wound healed, the second beast was coming forth.

CASE HISTORY FIVE

New York City, 1923–31

NAME: Rosie

PHYSICIAN: Multiple

CHAPTER 15

Madness

One would prefer a physical trouble which would produce outspoken feeble-mindedness with its limited range of harmful effects to this encephalitis which may produce an intellectual, tormented and cruel monster out of a gentle girl or boy.

Before long, the mild references in the British press to "altered dispositions" in cases of encephalitis lethargica gave way to new, disturbing descriptions: gross mental defects, homicidal attacks, change in moral character, loss of self-control . . . effects most often being seen in children. Another article went on to say, "Thus it is possible that some instances of motiveless crime and of violent behavior are the result of this disease, and not as might be supposed, of inherent vice."

Von Economo had started seeing these cases a few years after the initial outbreak in Vienna; they appeared in London and the countryside soon after that. One patient told von Economo that she was pregnant by the Lord and accused von Economo of being God himself. Another young patient had to be restrained as he was brought into the clinic, twisting, screaming, and begging the doctors not to burn him.

Reports were now surfacing of postencephalitic children who

attacked their family members, tried to murder siblings, or attempted to rape other children. One child tried to bite the penis off another; a female patient attempted to commit suicide by breaking her neck beneath the foot of her bed. Another child put his head in the fire on one occasion and took a hatchet to his sister on another. Still another nearly strangled a woman on the street. Children were known to smear the walls of a room with feces. They ran off rooftops, tried to hang themselves, jumped into rivers.

Doctors studying one group of encephalitis lethargica children in Philadelphia asked them to record their dreams. One drew a picture of the doctor and penciled flames bursting from his head. Another wrote, "A mother had a little girl and a man came along and took her for a ride. He then told her to shut up. She didn't. He put his hands over her mouth and cut her up."

What's worse, the children knew what they were doing. Numerous doctors reported that the children were not insane; they were completely aware, even afraid of their actions. That was one way to distinguish chronic encephalitis from other conditions like schizophrenia and psychosis. A truly insane person doesn't know he or she is insane; a victim of chronic epidemic encephalitis knows all too well what is wrong. Some of these children even asked to be restrained to keep from harming other people.

A patient herself gave the clearest description of what it was like: "It's so sad to be like me. This is only the beginning; it's going to get worse. You don't understand how it is not to be yourself. I feel so vicious at times. I was always good and kind to people. There are other people in the world like me. I feel sorry for them. I know a little girl like me, and I only pray that something will happen to her before she grows up. I want to tell you about this because the time is coming when I won't be able to. But you're well, you can't understand!"

When acute cases of epidemic encephalitis first broke out in New York, the concern was for immediate care of the patients—

whether in their home or in the hospital. From there, many went to the morgue. But others survived, and they were appearing in physicians' offices showing chronic symptoms. Already in England, the Mental Deficiency Act had been amended. Originally intended to provide care for infants born mentally incompetent, it was changed in 1927 to include the survivors of sleeping sickness.

The children of epidemic encephalitis presented a unique problem for public health—they were too young to be housed with adult patients and, by the very nature of their behavioral disorders, they were disruptive to the other patients. In the city, children with epidemic encephalitis were becoming an urgent problem—there was no place for them. They could not be kept in the hospital indefinitely, and they could not live in adult asylums. They were a danger to their own family members and strangers alike. Where do you put children such as those—children who had lost everything but the knowledge of that fact itself?

Tilney would later say of the sleeping sickness epidemic, "Old and young alike fell before it, but the most pathetic victims were the little children who survived its ravages. The disease put a blight on these poor children even before they had a chance to get started in life. They were not only handicapped by convulsions and physical ailments of various kinds, but their entire conduct underwent a serious change. Children who were lovable and tractable became just the reverse. They became a serious problem."

Parents at first struggled to keep them at home. Numerous physicians treated them and specialists were consulted, including Jelliffe and Tilney. But eventually the children's nature—horribly damaged by the encephalitis—was not fit for society. They stole, snuck out of the house to wander the streets, exploded in fits of rage, and tried to hurt others. For New York families who could afford it, these children found a new home, far away from the city, in farm asylums.

* * *

The first thing visitors would have noticed as they drove into the Kings Park State Hospital was the long, tree-lined avenue carving a meandering, manicured path toward Long Island Sound.

The next thing noticeable to any visitor was how much the asylum *looked* like a small town or village. There were no towering buildings with barred windows; those would come later. Instead, white, two-story clapboard cottages spotted the serene landscape of pines and oaks, pastures and wood trails.

On one side of the avenue was a fire station, an open field, and, in the distance, the stately houses where some of the facility's physicians lived. On the other was a series of buildings, a barnyard, and a smokestack and apple orchard in the distance. At its height, Kings Park Hospital occupied eight hundred acres. While cities modernized and moved away from rural life, farm asylums maintained a sort of old-fashioned existence.

When the train came into the town, the engine was cut and it coasted into the Kings Park station carrying coal, molasses, coffee, tea, and spices. Workers unloaded the supplies into the storage rooms of the depot, laying them out on long, butcher-block tables to be delivered to the kitchen house.

There was nothing for the train to carry back out of Kings Park. The self-sustaining community produced no waste, no trash—milk bottles were used, then left out for the milkman to refill; clothes were recycled and woven again; food scraps and kitchen garbage were fed to the animals; even laundry was done right there on the grounds as opposed to sending it to town for cleaning.

In the spring, Kings Park came to life as leaves unfurled and fruit trees blossomed. In summer, the brush and forests grew thick, and the white sails of boats appeared on Long Island Sound. Autumn brought butchering day, as well as the scent of sliced apples drying in sunlight and cider boiling. And in winter, barren tree trunks and bushes sat in the yellowing grass. The cottages and buildings looked naked without the lush greenery that softened them during the spring, summer, and fall. There was an unavoidable sense of dormancy, even death that accompanied winter there.

The Kings Park asylum had been opened in Suffolk County in 1885 on a lip of land that jutted out into the sound. Long Island had six farm asylums in all—each with several hundred acres. The farm asylums served as self-sustaining communities, like towns in their own right, where the patients could walk through the pastures and tree-lined roads. Above all else, there were no bars on the windows. This was meant to be a refuge, not a prison.

In addition to the theory that a restful respite from city life would do wonders for the nerves, it was believed that structure and responsibility were essential as well. So the patients who lived in these farm asylums were a working part of the society, with jobs in farming, gathering food, cleaning, gardening, laundry, etc.

Physicians regularly visited these farm asylums from the city or lived on the grounds themselves. Their focus was not to treat or cure these patients, but to establish a personal relationship in which the patient would trust and confide in the physician—in a sense, they were sowing the seeds of psychotherapy.

Occasionally, Kings Park made news. Fires were frequent there, as they were elsewhere, in the age of wooden buildings and questionable wiring. Other times, patients escaped. One of the more sensational cases out of Kings Park was the story of the Brush sisters. *Both* women had been committed involuntarily, and their vast personal wealth became the property of their doctors. Ten years later, they won a lawsuit against the doctors for false certificates of lunacy.

For all their later faults, at the turn of the twentieth century, farm asylums were the progressive answer to the care and health of the mentally ill. In 1920, asylums throughout the United States, Kings Park included, began seeing patients of a different kind arrive: children.

Insanity was considered an adult's disease, so few accommodations existed for children suffering the same. And sharing the same wards was, as one benefactor said, "undesirable for the adult and frequently ruinous for the child." Kings Park became the only state-sanctioned asylum to help care for the child victims of epidemic encephalitis.

In the overcrowded city hospitals, children were often confused with psychopathic children, but the doctors and nurses at Kings Park noticed a few marked differences among the postencephalitic children: "But as compared to the psychopathic child their behavior is more simple, open, impulsive and without malice, cunning, or regard for consequences. . . . These children have few friends and seldom belong to gangs, but are avoided and called crazy by their fellows." When the children first arrived, the difference between the two was immediately apparent. Whereas a psychopathic child is withdrawn, quiet, and slow in getting acquainted with people, the postencephalitic child shows a "marvelous ability to remember names, and is in close contact with the life about him."

If psychopaths and postencephalitics were easily defined, not all mental illnesses were. One disease encephalitis lethargica may have been confused with was schizophrenia. One modern medical historian, Mary Boyle, believes that in the confusing, blurred lines between psychiatry and neurology during the 1920s, many patients with behavior disorders were seen by psychiatrists, not neurologists. As a result, Boyle believes a number of cases diagnosed as schizophrenia were actually survivors of encephalitis lethargica.

The tragic distinction among these children was that they were completely unfit for society—but they were not insane. They lived in a world where terror came not from outside dangers, but from within their own mind.

Madness is as old as time. Archaeologists have located skulls dating as far back as 5000 B.C. with holes bored through them to allow the evil spirits to escape. In the Bible, madness was punishment from God. India's Hindus blamed madness on a dog demon, planting the idea of werewolves and the expression "black dog" for depression. Greek myths described insanity, and Shakespeare wrote about madness in twenty of his thirty-eight plays.

As long as there has been madness, there has been the struggle to handle it.

Even into the nineteenth century, insane family members could be found chained to the wall or caged in a hole in the ground. These "fools" or "village idiots" were the responsibility of the family, and the family was not always compassionate. One sixteen-year-old boy, found in a pigpen, had been there so long that he had lost the use of his limbs and had to lap his food from a bowl. And it wasn't only children at the mercy of their families. One wife chained her husband to the wall until his legs withered.

Obviously, the church or the government needed to provide help for some of these cases, and the idea of a hospital for the insane was born. The first was actually built by a religious order outside of London in the thirteenth century; it was called Bethlehem. Over time, people rushed the pronunciation until it became "*Bedlam*." Under Henry VIII, when most of those benevolent monasteries were dissolved by the laws separating church and state, the mentally ill became the wards of the state. They also became tourist attractions—sort of a human zoo—and they were especially popular on weekends and holidays. Londoners paid for tickets, came drunk, threw food, and sometimes assaulted patients.

Private madhouses began sprouting up throughout the country, and when a family could afford it, that was certainly the better alternative. It was in this era that the hospitals for the insane first began across the pond. Colonial America built small dwellings in the middle of town to confine the mentally ill, and as one historian wrote, in that sense, "confinement" goes back as far as colonial times in the United States.

By the early 1800s, the United States had only two hospitals for the mentally ill, one in Pennsylvania, one in New York. The New York Hospital opened the first psychiatric building designated a "lunatic asylum."

During the Enlightenment, physicians began a movement toward a more compassionate and practical solution, but the sheer numbers of the insane made that difficult, if not impossible. Toward the end of the nineteenth century, there had been a sharp rise in insanity, one that most likely emerged from a reclassification of diseases. Syphilis, a common plague of the time, led to neurosis. Alcoholism was considered a mental illness, and during that time period it seemed almost epidemic thanks to unsafe drinking water, which led people to drink ale or wine, whose processing killed bacteria. With no retirement or nursing homes, the elderly were crowded into the class of mentally ill as well. And the industrial age had created an atmosphere of less sunlight and poor nutrition. The result was a condition known as rickets, a softening of the bone, which made childbirth more difficult. So, birth injuries added infants and children to the long list of the mentally impaired.

Cities all over the world built grand hospitals to house these patients, a sanctimonious gesture by enlightened societies who at last believed the mentally ill were in need of decent care, and the Industrial Revolution allowed for these mammoth hospitals to be built. Medicine, too, acknowledged the mentally ill, even hoping to cure them in some cases—although often the cure was far worse than the disorder.

As the world became more civilized, society more sympathetic, and medicine more progressive, mental hospitals of a softer kind came into vogue. Asylums were first intended to be just that—an "asylum" from the hectic, fast-paced world of the city. Asylums would be places in the countryside where people who suffered from frazzled nerves could recover in peace; they were sanitariums where sanity could be restored.

In the United States, the person who led the fight to reform treatment of the mentally ill and to develop asylums was Dorothea Dix. Often neglected in history, Dix was a nurse who was teaching a Sunday school class at a local prison in 1841 when she noticed mentally ill patients chained to the walls. She fought for government intervention in treatment of the mentally ill—at the time, a radical idea. She even proposed legislation that the federal government set aside several million acres for asylums—the bill was passed by both houses, but vetoed by President Franklin Pierce.

City hospitals in America and elsewhere were brimming with patients who needed long-term care, and the idea of the farm asylum materialized. Toward the end of the nineteenth century, these farm asylums and sanitariums appeared throughout the world. But, as idyllic as they sound, farm asylums began showing their cracks shortly after World War I. More and more buildings had to be built to accommodate patients. More and more nurses and physicians were needed to treat the growing number of patients. "The reformers were defeated," one historian noted, "not by the faulty nature of their concept but by the pressure of numbers."

Part of the problem was the very nature of mental illness—chronic. Some of those patients would live out their lives in these asylums, without hope of a cure. One neurologist working in a New York City asylum complained that there were three hundred patients per doctor. It was also not a field many doctors were drawn to—it was dark, depressing, and exasperating work with little hope of patient recovery.

During the 1920s, however, when the farm asylums were still functioning well and had not yet turned into the warehouses for the insane they would one day become, they offered a humane alternative, at least for a while, to the children who had survived the sleeping sickness epidemic.

* * *

In 1924, Kings Park built a colony for these children. Although the cottages were intended for any juvenile with mental illness, it was epidemic encephalitis that prompted the building of the colony. Of the first fifty-eight patients, forty-one were survivors of the encephalitis lethargica epidemic. As soon as the colony opened, children arrived from hospitals in Manhattan, Brooklyn, and elsewhere on Long Island.

The cottages were located along the main, tree-lined avenue beside Wisteria Hall, amusingly called Hysteria Hall by the staff. To the back of the colony were the train station, the apple orchard, and an engineer's shed. The cottages were meant to be as much like a home as possible. They were two-story, redbrick buildings that had been painted white with dormer windows—without bars. They featured day rooms and screened porches for summer and several fireplaces for winter. There was even a metal swing set outside.

From the first, the children were a challenge for hospital staff. Children with postencephalitic problems were remarkably similar in their behavior, marked by emotional instability, quarrelsomeness, irritability, tantrums, and breathing tics. The aim for the hospital staff was to retrain these children to show emotional control and social adjustment, and in the best cases, they hoped to be able to send children home. The first days, according to one occupational therapist, were the worst: "Days of fighting, biting, scratching, lying and tale telling . . . These were hourly or half-hourly events." The staff was close to tears from frustration and exhaustion. The therapist added, "There seemed to be nothing normal about these children. They either grabbed all food in sight or wanted no food at all; they slept half the day and stayed awake half the night; homosexual practices and masturbation were prevalent with both boys and girls."

The staff constantly reminded themselves that they were dealing with manifestations of a disease and not wanton misbehavior. Their first glimmer of hope was that the children themselves seemed to want to do better.

The children adhered to a strict schedule:

6:30 A.M.	Rising hour. Toilet, teach patients to clean teeth and to use mouthwash. Teach patients to dress properly, taking one article of dress at a time. Pride in personal appearance must be stimulated.
7:15 A.M.	Breakfast. Oversight should be given to the consumption of a sufficient quantity of food.
7:45 A.M.	Patients should be taught to make their own beds, possibly working in pairs. This will tend to stimulate teamwork.
9:00–10:30 A.M.	Simple academic upgraded schoolwork.
10:30 A.M.	Toilet, prepare for outdoors. Glass of milk.
10:45–11:45 A.M.	Exercise, games, marching, etc., out of doors if weather permits, open windows if indoors.
11:45 A.M.	Toilet, prepare for dinner.
12:00 noon	Dinner.
12:30–1:15 P.M.	Clear up dishes, put dining room in order, rest.
1:15 P.M.	Toilet, drink water, prepare for class.
1:30–3:15 P.M.	Simple handwash; last part of period may be used for preparation of songs, recitations, fancy dancing for future parties and entertainments.
3:15 P.M.	Toilet, prepare for outdoors, glass of milk.

3:30–4:45 P.M.	Both groups out of doors, walks, games, exercises, etc.
4:45 P.M.	Toilet, prepare for supper.
5:00 P.M.	Supper.
6:00–7:30 P.M.	Quiet games and reading aloud.
7:30 P.M.	Prepare for bed, clean teeth, bathe, toilet.

Added to the schedule were weekly plans for singing lessons, marching, games, and dancing. On Wednesdays they watched moving pictures, and on Saturdays during the summer months, they played baseball.

In all, there were two cottages, one designated for the girls and one for the boys. Inside, their home life was to include the daily chores any child would have. With so many children under the age of sixteen, the vast majority of whom had their intellect still very much intact, the nurses set up a schoolroom to educate the children as well. When they did misbehave, small privileges were revoked.

The results were remarkable. The children showed healthy appetites, good manners, improved personal appearance, and evident physical improvement. The nurse wrote, "This training has been so well handled and the granting and withholding of small privileges so effective in encouraging the children to greater efforts at self-control that the problem of their management has been materially reduced. . . . Emotional stability is not entirely established but they show themselves susceptible to training and there are now many instances of great self-control."

Over the next five years, 114 children lived in those cottages at Kings Park. And they would soon serve as guinea pigs in the vaccine trials in the coming decade. After all, without hope of a vaccine or cure, the children had little or no future ahead of them. Those children at Kings Park proved to be the lucky ones. For thousands of other children damaged by epidemic encephalitis, there was no asylum, no serene solution to their madness. Many ended up in prison or juvenile homes. In a sense, they did die of acute encephalitis lethargica. Their bodies remained, like a ghostly shape of the children they once were, but their minds were horrifically changed. As a British doctor explained, "When the child's personality changes so dramatically, they've 'lost' the original child forever."

Neuropsychiatrists in New York had been so focused on the epidemic of encephalitis itself—studying it, diagnosing it, treating it—they were unprepared for the blow it dealt them *after* the epidemic had subsided. Jelliffe had already experienced psychological changes in his patient Adam. Tilney, too, had started seeing patients, children, affected by the epidemics.

On a cold December night, both doctors made their way up Fifth Avenue to the New York Academy of Medicine to hear a lecture on that very subject. Nothing could have prepared the physicians for the case they would hear about that night.

CHAPTER 16

Rosie

Jelliffe and Tilney sat in the deep red, velvet chairs of the auditorium. Outside, it was cold, and frost powdered the front steps of the building of the New York Academy of Medicine. Across the street, in the early moonlight, the tree branches of Central Park reached like talons toward the nighttime sky.

The footsteps could still be heard in the chilly atrium, as well as the muffled voices. Overcoats, hats, and gloves were checked; snow was kicked off shoe heels. The physicians filed down the stretch of red carpet into the auditorium for the meeting of the New York Neurological Society.

Tilney had served as president a decade before this night. Both he and Jelliffe had been on the main council of the society for years. Tonight, they sat in a room smelling of damp wool suits, tobacco smoke, and musty seats to hear the president of the society give his lecture. S. P. Goodhart, also a longtime member and someone who had served for years with Tilney and Jelliffe, was giving a talk about an unusual patient of his. This was not the first time epidemic encephalitis had been brought before the society. As early as 1919, it appeared in the minutes of the group's meetings. Following that, it had appeared intermittently in the programs. An entire evening had been devoted to the subject.

On this cold night, however, the tone would be different. All of the neurologists present, more than one hundred of them, had most likely seen cases of epidemic encephalitis. Many had watched the long, silent slumber. Some had seen the chronic symptoms beginning to emerge. None had seen a case like this one. Dr. Goodhart climbed the steps to the stage and stood before the lectern to tell the tragic story of Rosie.

It is not known which doctor at the Neurological Institute originally saw Rosie; it could have been any neurologist on the ward that day. What the doctor would have seen was a pretty girl, her dark hair cut into a pageboy, with a frame as frail as bird bones. Her thin legs turned in slightly and her shoulders slumped forward. She looked, as many teenage girls do, caught in that lapse between childhood and womanhood, unsure of what to do with her new, longer limbs.

She had actually been seen in the institute five years before when she was just a child recovering from a February case of the flu. It was not pandemic flu, just a normal strain that circulated, and still does, each winter. Her record filed at the institute was vague, denoting fever. A diagnosis of her case was never clearly noted in the hospital records. Her case report, given by her mother, described a very normal and healthy child. She had the usual bouts of measles and chicken pox, the usual bumps and bruises as a child. She started school at the appropriate age and was described as well liked, sociable, and clever. Little else is known about her personal life. Her mother appeared to the doctors to be an intelligent, concerned parent.

The only unusual thing in her case history was the fact that her eyes had grown weaker, causing her to have trouble reading, her left arm had drawn up against her chest, and she had started walking with a limp. To Rosie's family, the symptoms were a nuisance; to her neurologists, they were indicative of a breakdown within the brain. In such a complicated wiring system, one failure in the brain will likely have a domino effect. Rosie's symptoms signaled that the dominoes were beginning to fall; what the doctors didn't know was how ruinous the collapse would be.

By the next year, Rosie's physical complications were growing more apparent, and she was taken to Mount Sinai Hospital, where she was diagnosed with Parkinsonism, a diagnosis fast becoming the norm for most postencephalitics. It was also noted that her pupils were unequal. Though both responded to light, the right was larger than the left. Her mother complained that she had also been unusually tired, falling asleep in class and having trouble with her schoolwork. The young girl was discharged from the large hospital opposite Central Park and walked out into a winter day. She would not return to a hospital for another four years.

Rosie was again taken to the Neurological Institute, this time in 1928. Her case history, accurate and utterly impersonal, recorded a masklike expression on her face, a spastic gait, a left arm flexed with a significant tremor, a squint in her left eye, and a deviation of the tongue to the left. Her blood and spinal fluid tests failed to detect anything out of the ordinary. She was given a diagnosis of "chronic epidemic encephalitis," which would have been the logical cause of her Parkinsonism.

By then, Rosie was in the eighth grade, and she returned to school after her visit to the Neurological Institute, but she struggled, unable to concentrate and frustrated by eye spasms that made reading difficult. It was during this time that the first sign of a personality change became obvious to her family. Her temper became unmanageable, and Rosie tore at her clothing and attacked her mother.

In shock, her mother waited for her to calm down and asked, wide eyed, what was wrong with her.

"I couldn't help it," Rosie said sincerely.

She also could not help the subsequent explosions when she broke windows, threw tantrums, and lashed out at her sisters, each time growing sleepy and lethargic afterward. In a dazed trance, Rosie would repeat, "Why do I do it? Why do I do it? I can't help it."

It was during this time that Rosie began locking herself in the bathroom. The first time, she reappeared sometime later, trailing blood and missing teeth. Horrified, her mother demanded to know what had happened. Rosie calmly told her that she "could not help taking them out."

Rosie would remain awake, anxious and irritated, waiting for family members to go to bed. Once they did, she returned to the bathroom, shut the door, and began extracting her teeth again. She pulled all but nine of her teeth. An infection forced Rosie back into the hospital in March—this time Bellevue. A dentist there had to remove seven more teeth due to infection and damage.

The next few months were relatively quiet for Rosie. Spring came and went, and summer now slouched around the city buildings, leaving shards of sunlight in places and pools of shade in others. Heat wilted the leaves on trees and warmed the pavement. Horse flies, heavy and slow, hung on the hot air. Rosie's mind, like summer itself, began to seethe.

Self-mutilation, as Rosie's case developed into, was not an anomaly. History is full of examples of self-injury ranging from mild to extreme. Self-flagellation was practiced among religious groups. A late-nineteenth-century study looked at a number of women in Europe, known as "needle girls," who pierced their own skin with sewing needles. Self-injury was also seen by physicians during World War I when soldiers would shoot their own hands or feet in order to escape certain death on the battlefield. And teeth in particular seemed tempting to self-harming patients. Given the time period and poor dental health that may not be surprising—epidemic encephalitis was also known to cause nerve damage and rotten teeth, which may have prompted Rosie to remove hers. But one cannot overlook the motivation behind the behavior either.

Interestingly, self-injury is also a twentieth- and twenty-first-century problem. People who practice it search for a way to stop, if only momentarily, the internal pain with an external one. They describe a sense of immense relief when inflicting pain. Researchers have found that self-injury occurs most often in girls and tends to peak around the age of fifteen, declining after the age of sixteen. Rosie was not yet that age, but she soon would be.

While Rosie's case was certainly not in the mainstream, self-mutilation was something physicians had dealt with for centuries. There is even a martyred saint for its cause. Saint Lucia, who is celebrated in December when there is the least amount of sunlight and the longest hours of darkness, was a young woman relentlessly pursued by a pagan suitor who told her she had beautiful eyes. A devout Christian, she plucked out her own eyes and sent them to him with a note asking to be left alone from that point forward. It is said that when the martyred Lucia arrived in heaven, God restored her vision and gave her an even more beautiful pair of eyes.

Self-injury can be found in everything from medical texts to Shakespeare to the Bible: "And if thine right eye causeth thee to stumble, pluck it out and cast it from thee."

It was a hot July when Rosie's mother admitted her to the newly opened Morrisania Hospital. The Morrisania complex was built in the Bronx as part of a revitalization effort in that part of the city and had opened just two years before. The hospital was impressive, designed by the same architect who designed the old Tammany Hall headquarters, the New York City Courts building, and the Tombs. Rich in architecture, with a red tile roof, sand-colored bricks, and an ornate, Neo-Renaissance façade, the hospital was also well lit with sun parlors, a courtyard, and a garden.

It had been months since Rosie's last trip to the Neurological Institute, but her mother noticed that one of her eyes was swelling and reddening. She was also running a low temperature of one hundred degrees.

That month, there was a record heat wave, spreading across the Midwest, moving sluggishly toward New York, and finally reaching the city in the last week. Already, the heat wave had claimed eighty lives nationwide. In New York, on July 30, it was the fourth consecutive day of unbearable heat tipping the thermometer at ninety-five degrees. At the hospital, fans whirred, and windows were opened for the breeze. One of the house physicians examined Rosie, admitted her to a bed, and washed her eye with boric acid. An ice pack was placed over her eye. Outside her window, in the distance, a weak wind moved across the shimmering Harlem River, turning the surface of the water reptilian.

There was also a full moon on the last days of July, brightening hospital rooms late at night. Science has never been able to explain the influence of a full moon. It controls ocean tides. Studies have shown more violent acts occur under a full moon. More pregnant women go into labor during a full moon. More people are admitted to the hospital. And there has long been a belief that a full moon gives rise to its namesake: lunacy.

Rosie lay in the hot hospital room in a gown as soft as a cobweb. She was lulled into a quiet sleep, still hearing the fans, the commotion in the hospital halls, and voices outside her door.

Occasionally, a radio's hollow sound echoed through the hall broadcasting news about the heat wave, as well as a plague of grasshoppers that had destroyed so many crops that the federal government was called to action. Even the newspaper headlines bemoaned in biblical tones the high temperatures, "leaving in their wake a long trail of death, suffering and destruction from pests and fire."

That night as Rosie tried to sleep, a rumbling thunderstorm moved into the city, with winds clocking over forty-five miles per hour and carrying the rain that would bring an end to the blistering heat wave. On that last day of July, at 3:40 A.M., as the storm was coming to an end, the winds quieted, the air cooled, and the last of the rain streaked the windows, an attendant went to check on Rosie.

The woman came screaming out of the room.

The attendant hurried to the floor desk to find the nurse in charge of that ward. The nurse rushed into Rosie's hospital room and found her sitting calmly in bed, holding her right eye in her hand.

The nurse asked Rosie what had happened, and Rosie, completely stoic, told her that it had just fallen out while she was sleeping. The nurse later reported that Rosie's reactions were quite normal except for her seeming indifference toward what had happened. She did not complain of being in any pain, and she did not seem in the least disturbed. Somehow, wrote her physician, "the encephalitic process affected the pain perception mechanism so that its activity was suspended for a brief interval during which time this remarkable self-injury took place."

The next morning, the storm gone, the light was liquid, pouring through the hospital windows, seeping down the hallways. There is no record of which room Rosie stayed in, but given the hospital's well-designed windows and views, her room was likely filling with sunlight by 7:00 A.M. when Rosie was found lying in her bed. Her left eye looked normal and the socket of the right one had been dressed with gauze.

The neighborhood around the hospital was coming to life, and doctors and nurses made rounds through the ward. Milk bottles had been delivered quickly, packed in thick wooden boxes filled with ice to keep the milk from spoiling. Canvas awnings unfurled over storefronts, and newsboys untied the morning papers. The metallic rattling of the Third Avenue el could be heard. The whole block had an unusual, earthy, dank smell from, of all things, mushrooms. With Prohibition in full force, many of the breweries concentrated in the Morrisania neighborhood were left with empty caves that had once kept beer cool in the summer. To make ends meet, they used the caves to grow mushrooms in compost.

Stronger even than the compost were the smells of chicory coffee, eggs, and bacon that wafted from the hospital kitchen. Somewhere in the courtyard, through the open windows, cigarette smoke floated in the undercurrent of morning scents. The rooms were filling with short-lived light. Soon, the drapes would drop or shutters would close, casting all of the rooms into blue-gray light during the hottest hours. Patients were just waking, and visitors were beginning to make their way into the hospital. Every so often, the scent of fresh flowers, already beginning to wilt, was carried in the breeze from the hall.

The calm of the morning did not last.

At 8:45 A.M., shouting was heard in the ward. A nurse ran toward Rosie's room, her shoes clapping against the tile floors. When she opened the door, there was very little blood in the grisly scene so vibrant red was not what first attracted the attention of the nurse. Rosie, lying calmly in the bed, told her that her *other* eye had fallen out. The nurse came to her bedside and began frantically pulling at the blankets and bedsheets, while Rosie's hands groped blindly along the bedding. Finally, they found the missing eye, tangled in the sheets of the hospital linens. Again, Rosie seemed to be in no pain and was completely at ease.

The next day, Dr. Goodhart was brought in to examine Rosie.

She sat before him as a waifish sixteen-year-old girl with no eyes and missing all of her teeth. Goodhart found Rosie to be unemotional, cooperative, well oriented, and of normal intellectual capacity. She denied having injured herself and could not remember the details of why her eyes had "popped out." An X-ray examination of her head showed two dark orbits on an otherwise normal skull. An ophthalmologist saw the girl and noted that her eyes had been removed almost perfectly, with the fibers and muscles still cleanly attached.

Rosie showed a tendency toward more self-harm, scratching at her cheek and biting deeply into her tongue, and by October, she was admitted to Montefiore hospital for children. During her first few weeks, she picked at the dressings over her eye sockets and rubbed her face repeatedly. She was then restrained and watched carefully for a full month. No further self-injuring behavior occurred.

Hospital staff found her to be very cooperative and tidy. Her stream of thought was normal, with no hallucinations or delusions. She was occasionally irritated with the nursing staff, but aside from that, Rosie was very alert and "showed a lively sense of humor." In general, she showed no depression, weeping, or suicidal tendencies, but would grow very sad when the self-mutilation incident was brought up. She insisted that she did not remember what happened to her eyes, but she did recall pulling her teeth out and the peculiar force that compelled her to do those "horrible things." Goodhart suspected otherwise; amnesia had proved extremely rare in the encephalitis lethargica cases.

Rosie improved both physically and mentally, even learning Braille reading during her stay at the hospital; but it would take another four months for Rosie to admit to Goodhart that she had actually torn her eyes out with her own fingers. Again, she shrugged it off as an irresistible urge.

"I was like hypnotized at the time, something made me do it," she said by way of explanation.

* * *

It was three months later that Dr. Goodhart stood before Tilney, Jelliffe, and hundreds of other neurologists to give his lecture, "Self-Mutilation in Chronic Encephalitis: Avulsion of Both Eyeballs and Extraction of Teeth." His article was published in the *American Journal of the Medical Sciences*. He also decided to film some of his sessions with Rosie. It was a new form of documentation that doctors found particularly helpful. Tilney had also filmed patients, so that in slow frame, he could watch exactly how his Parkinson's patients were moving. Goodhart included in his film a shot of Rosie's eyeballs displayed on a cloth.

Still, no record of what ultimately happened to Rosie exists, and Goodhart's article on the subject, published that same year, gave no indication. Like so many others, the end of her case history was either forgotten or destroyed. For many patients like Rosie, the insanity caused by epidemic encephalitis was overshadowed by the physical disabilities. Over time, the inevitable Parkinsonism set in. They became living statues confined to wheelchairs or beds. If there was still rage within them, it was buried deep beneath the stony surface.

Given the fact that she was rapidly improving in 1932, Rosie may have been allowed to go home and attempt a life more like the one she could have had before epidemic encephalitis.

If not, she would have lived the rest of her life in an asylum—in the same condition or worse until the day she died. Perhaps it was then, after her death, that Rosie was given a more beautiful pair of eyes.

CHAPTER 17

The Neurological Institute

The *Olympic* sailed across a gunmetal gray sea rife with storms, finally arriving in the glassy harbor in autumn of 1929 with a clear view of New York's skyline. Rather than a rocky island weighed down by buildings, Manhattan looked more like a city carved from a mountain range, molded and sculpted into geometrical shapes. If most cities adapt to their terrain, it was just the reverse in New York. And at night, from a distance, the building shapes dimmed, leaving only rectangular window light and a skyline studded with honey amber.

It was a view that had inspired many. In *The Great Gatsby*, Fitzgerald described a view of the city seen for the first time as one of "wild promise of all the mystery and the beauty in the world." Ezra Pound described the great buildings as magical, "squares and squares of flame, set up and cut into the aether." Ayn Rand saw the skyline not in detail, but in shapes: "The sky over New York and the will of man made visible." And Frank Lloyd Wright saw it through an architect's eyes: "The buildings are shimmering verticality, a gossamer veil, a festive scene-prop hanging there against the black sky to dazzle, entertain, amaze."

Constantin von Economo had not been to New York in thirty years, and the city was almost unrecognizable. From the ship, the skyline seemed shrouded in sheer clouds, but even through the haze, the buildings rose high above the gauzy air. Amid the titanic structures, crowded streets, and glowing lights, the von Economos could not help but feel "enchanted," his wife later wrote.

Before making their journey to America, the couple had taken a trip to Bern, where von Economo worked with European and American neurologists to organize the International Neurologic Congress. Von Economo's wife later noted in a book she wrote about her husband that, since the war, von Economo had been instrumental in reconciling the broken relations between German and French neurologists. At long last, Europe's neurological community was coming together, though it was still far behind the advances being made in America. Von Economo and his wife had been invited to attend the grand opening of a huge new Neurological Institute in New York that fall; he would also be giving a talk to the New York Neurological Society. And one person sure to be in the audience was S. E. Jelliffe, who once said of von Economo, "not only was he a scientist in the best sense of the word but peerless as a man."

In an apt analogy between von Economo and the city of New York, Jelliffe wrote, "Specialist medical towers go up like the sky-scrapers almost unperceived by the busy hard-working physician. Occasionally, however, there occurs a growth roughly comparable to the Rockefeller Center, which no one can fail to see. . . . Such a development took place in the field of neurology when in 1916–17 Constantin von Economo of Vienna made his observations and published his work on sleeping sickness."

Von Economo envied the American scientists, who were not burdened with federal control of clinics. In postwar Europe, von

Economo had continued his work, publishing an impressive number of papers. But Europe had fallen into a recession after the war, and as a government program, scientific study was falling to the wayside.

He and his wife walked along the streets, beneath the shadows of buildings and scaffoldings crossing the pavement. They passed beneath the giant frameworks of steel bars and reinforcements for the raised tracks, bridges, and highways. In spite of progress in every direction, some problems still existed, including sanitation. The city's Public Health Committee had recently estimated that the regular tonnage of ashes, garbage, dead animals, street sweepings, and rubbish piled together in the city would be higher than the Woolworth Building. The *New York Times* published an article lamenting that a "New York as clean as Havana is still a dream." It must have created a strange and malodorous contrast—the clean, elegant, streamlined modern architecture fronted by chaotic piles of crates, loose papers, spoiled food, and burlap sacks. For the von Economos, however, the mounds of refuse were nothing compared to the titanic buildings.

Throngs of people and constant noise filled the streets, but it was the sound of promise—metal against metal, hammering, rumbling subways, automobiles, and train whistles. The wealth of New York was obvious in the architecture, the size and the sheer number of buildings. By 1929, 709 new buildings were planned. Between 1928 and 1931 alone, the Chrysler Building, the Bank of Manhattan Company, and the Empire State Building were erected—all three competing to climb higher into the clouds than any other in the world. The buildings, gigantic in scope and material, somehow appeared weightless, suspended in the sky, literally piercing the clouds on some days.

In addition to those massive skyscrapers, there was the building of major hospital complexes, the West Side Highway was under construction, the Holland Tunnel had just opened, and the George Washington Bridge was midproject. Everything about New York, from its architecture to its skyline to its medical facilities, felt thoroughly "modern."

Standing on a New York street, staring skyward at the tallest buildings the world had ever seen, must have made it hard to believe the economic foundation beneath them was beginning to quake.

In spite of all the building and an unprecedented economic boom in recent years, Wall Street felt the first tremor of trouble in the summer of 1929 when sales had slowed down and unemployment began to rise. A sharp rise in the Dow Jones that August reassured people—and blinded them to the inevitable fall. New York had seen a brief panic in 1907, which led to the formation of the Federal Reserve, and a recession in 1921, which had been a response to the war ending. The postwar recession was steep, but short. Soon after, returning soldiers entered the labor force. The factories that had been producing war materials now had an abundance of new technology and consumer products to build. Automobiles went from luxury items to mass-produced, affordable ones, a cultural shift that also changed the domestic gas and oil industries, which enjoyed a 250 percent increase in profits; home appliances were available; electricity and plumbing became commonplace; and there were radios to purchase. Consumerism was at an all-time high, and credit was cheap, so it wasn't long before the average American overbought and went into debt. In addition to that, a number of novice investors flocked to the New York Stock Exchange, the largest in the world, to invest their money. Others put it away in the safest place possible: a bank.

The crash of 1929 happened in fits and starts over a series of weeks when the stock market dropped, then recovered, then dropped again. J. P. Morgan & Co. stepped in, just as it had done in 1907, to pump money into the market and level it. In 1929, the rescue proved ineffective as Black Tuesday, when the Dow Jones fell to unparalleled depths, left Wall Street with a $30 billion loss by week's end. Contrary to popular belief, an initial run on the banks didn't happen, and there were not dozens of people jumping from windows, although there were a few notable ones. Winston Churchill witnessed a jumper outside his hotel window in New York, and the Statue of Liberty saw her first suicide in the spring before the crash. In 1930, the market saw an upswing, with a decline in prices and a steady hold on wages. In some sense, the year 1929, rolling into 1930, turned out to be a time of false confidence. It gave no indication of how bad things were about to get, but was instead the moment of inertia before the fall off a steep edge.

Historians have debated the actual cause of the 1929 crash, but most agree it was a complicated combination of factors like consumerism, market speculations, easy credit, and business monopolies. Only two hundred corporations owned almost half of U.S. business wealth. In the boom of the 1920s, the antitrust laws established in the Progressive Era grew lax. In such a top-heavy economy, it was bound to tip.

In spite of the brief economic respite in 1930, Americans were nervous at that point; things looked too uncertain. They stopped buying things. By the end of 1930, the drought that had started that summer—the one that would help create the Dust Bowl—was in full force. It crippled American agriculture and sent commodities plunging. At that point, there was finally a run on the banks, but it was too late. The banks themselves were closing. People lost their savings and their jobs, and they had no unemployment insurance. New York was hit harder than any other American city.

The economic downturn did not dull Tilney's vision of New York as a model for neurological advancement. Not even a crashing economy could slow the momentum of neurology at that point, and the importance of encephalitis lethargica in the pacing of that progress was evident in the medical literature of the time. As famed neurologist Bernard Sachs later wrote: "Encephalitis appearing in epidemic form had revolutionized the practice of neurology, so that in the future it would never quite be the same as hitherto."

Tilney's dream was looking more realistic by the time the state-of-the-art building for the Neurological Institute and its twin Psychiatric Institute, the country's first of its kind, were finished—they gave solid form to the vision Tilney had. With J. P. Morgan's contribution in honor of his wife, the Neurological Institute also had an immediate purpose to serve. When he toured the building, von Economo marveled at the fact that an entire floor of the institute, the Morgan Ward, was used for the study of encephalitis lethargica in New York.

In keeping with the changes in American medicine, hospitals like Bellevue and New York Hospital underwent a transformation as well. Columbia Hospital converged, combining the old Presbyterian Hospital, at Seventy-first and Park, with the Babies Hospital on Lexington and the College of Physicians and Surgeons in midtown. The complex had been built on Hilltop Park, the playing field of the New York Highlanders, which had changed its name to the Yankees years before.

The mammoth hospital complexes built during this time period physically reflected the changing focus of medicine, creating buildings for specific specialties. The new Neurological Institute was built as part of the Columbia complex at 168th Street and Fort Washington Avenue, and beside it was the newly opened Psychiatric Institute. When the Neurological Institute opened in 1929, it also took on another incredible challenge. With Tilney at the helm, Columbia's College of Physicians and Surgeons, the Neurological Institute, and the New York State Psychiatric Institute would combine to offer a new type of education that would include classes and hands-on demonstrations. In the past, psychiatric institutes were merely custodial, with very little research or study available. Tilney hoped the side-by-side institutes in Columbia's hospital complex would become the greatest place in the world for the study of the brain.

Tilney, in explaining the need for such a program, said: "The more highly socialized we are, the more complex our existence becomes." The life of prehistoric man, he said, had been simple and "far removed from the turmoil that follows in the wake of civilization" and "the machine age." Tilney considered it deplorable that one out of every twenty-five people suffered a nervous disorder. "The human race has not yet begun to recognize the brain as both an asset and a liability. . . . " Tilney's words, in fact the whole program itself, continued to knit neurology and psychiatry together. "We are today witnessing the dawning of a new era, when man is at last awakening to the importance of his brain and is seeking to understand it and its functions," said Tilney.

The institute was made to look as little like a hospital as possible. It was fourteen stories high, with the entire second floor devoted to doctors' offices, so they could be as close as possible to the surgical wards and patients. There were no glaring white walls; all hallways, wards, and rooms were painted in soft colors, with the operating rooms in a serene green shade. The new building held a large amphitheater with tiered seating for teaching courses, and world-renowned lecturers were invited to teach at the institute. Among the distinguished physicians and professors to lecture in the program would be Jelliffe, who was by then president of the American Neurological Association, a position bestowed upon him late in his career because of his unpopular loyalty to psychoanalysis. Jelliffe's lectures would cover the mental and nervous reactions to general disease—a subject he knew well in the midst of the encephalitis lethargica epidemic.

On December 5, von Economo gave his lecture, in English, at the Psychiatric Institute in the Columbia medical complex. When von Economo first began studying neuropsychiatry, only some 20 areas of the brain had been identified. By the time von Economo gave his lecture series in New York, he alone had isolated another 107. That evening, he lectured on the evolution of the brain and the obvious, superior intellect in humans, which he also considered to be highly hereditary. He also read an important paper about the part of the brain that controls sleep, a subject he had come to know well over the previous decade, and one that the New York neurologists present that night had grown more and more confounded by.

With impressive, world-class hospital buildings in place, two additional factors would help Tilney realize his vision for New York as the center of the world's brain study.

The team of New York neurologists did not yet know how world history would help accomplish their goal. Throughout the 1930s, especially after 1933, the Nazis gained popularity in Europe. With the Nazi rise to power, some of Germany's most valuable scientists, who were Jewish, migrated to the United States—New York in particular. For nearly a century, America had admired Germany's medical system—American schools like Johns Hopkins had emulated it; American physicians graduated from medical schools only to travel to Germany to learn from the masters; and American doctors often learned German in order to read the latest medical studies in their original language. Now, in a swift turn of events, America would replace Germany as the world's center of science.

The second stroke of luck for Tilney and the Neurological Institute was a man named William J. Matheson.

CASE HISTORY SIX

New York City, 1934–45

NAME: Sylvia

PHYSICIAN: Dr. Josephine Neal

CHAPTER 18

The Matheson Commission

It is strange that William Matheson loved the ocean. As children, he and his brother were sent overseas to Scotland to boarding school in the 1860s. En route, the schooner encountered a hurricane off the coast of Newfoundland and sprang a debilitating leak. The ship was demasted, and huge waves washed over the decks, taking the captain and a cabin boy overboard. The ship, gutted and wounded, floated through the sea with the remaining passengers. All food and fresh water supplies were lost in the storm, so someone aboard caught a porpoise, and they survived on the meat and salt-water coffee until being rescued by a passing ship called the *Marmion*. Matheson and his brother survived, but Matheson suffered from exposure and health problems well into his twenties. Decades later, Matheson's son would name his own yacht the *Marmion*.

Ironically, it would be aboard his yacht the *Seaforth* that William Matheson would eventually lose his life from a heart attack. Between his boyhood voyage on the *Marmion* that would save his life and that final sail on the *Seaforth*, Matheson survived a great deal and accomplished even more.

William John Matheson was not born into wealth, but a modest family of Scottish descent living in Wisconsin. During his schooling in Europe, he earned a degree in chemistry and subsequently began working with synthetic dyes. At that time, France and Germany produced the finest dyes, in all colors. Matheson formed his own chemical dye company in America, but kept several production plants in Germany as well for the higher quality. As Europe edged its way toward war, Matheson was all too aware that his supplies in Germany would soon be cut off. He became an expert in patent law, and as soon as war broke out, he closed his plants in Germany and seized the interwar opportunity to overturn German patents. Matheson began production of the same dyes in the United States and produced chemicals for the war effort.

Matheson's company occupied a large, multistory building in New York, and he amassed enormous wealth, with a city house on Park Avenue, a country home on Long Island, and a winter estate in Key Biscayne, Florida, which he named Swastika after the Indian word for sun. The name was changed during World War II. When Matheson died, the *New York Times* reported that his estate was worth $23 million, not including real estate or life insurance.

Matheson's health had always been questionable—from the childhood shipwreck that led to exposure and starvation to a prognosis, when he was thirty-five years old, that he had an enlarged heart and would not live another six months. Lack of sleep and poor nutrition during his service in the war further weakened his health. During the war, Matheson also contracted a case of the flu. It was shortly thereafter that he noticed his right leg began to tremble when he would stand; his left one would tremble when he sat.

Later, Matheson experienced periods of lethargy and weakness, sometimes running a low-grade fever with no discernible cause. It bothered him enough to cancel appointments and, even more telling, golf matches. Matheson began a frustrating round of tests and visits with a number of different physicians. Most doctors would not tell Matheson what his diagnosis was, and one told him it was "toxic palsy." Another doctor refused to give any diagnosis at all, claiming he had no idea what was wrong. Yet another doctor told Matheson that he had an illness that "came only to people of hardworking correct habits," and that it would be gone in a year's time. Still another doctor gave Matheson electric currents to his head, which Matheson found less than impressive: "I feel that this treatment produced no different effect than lying flat on my back would produce without the current."

Matheson spent the next nine months traveling around the world, and when he returned in 1922, his tremor was worse, he had trouble working, and he had an unusual gait. He tried a long list of treatments including hypnotics, glucose, radium water, cod-liver oil, diluted hydrochloric acid, ox gall, and Christian Science, among others.

In July 1922, he finally saw Dr. Charles Dana, a highly respected New York neurologist. Dana's *Textbook of Nervous Diseases* was already in its tenth edition, and his involvement in the New York Academy of Medicine and its Public Health Committee had solidified the relationship between neurology and the city health department even more. The health committee offered its official opinion on issues as varied as heroin addiction and daylight savings time and broadcasting health information by radio. Dana himself had been skeptical about an epidemic of encephalitis lethargica when it first broke out, saying in 1919, "encephalitis lethargica was exceedingly rare, probably not contagious, and certainly not epidemic." That same year, however,

Dana diagnosed a case of encephalitis lethargica himself and his opinion changed. In fact, Dana's first sleeping sickness case had originally been seen by Tilney, who misdiagnosed the disease. That mistake alone sheds light on the difficulty even within the neurological community in recognizing this disease. By then, Tilney had published his own book on the subject, and yet he missed this diagnosis.

It is also interesting to note that Dana did not inform his first few patients that they *had* epidemic encephalitis. Very much in keeping with medical thinking at the time, physicians sometimes considered it in the best interest of the patient to withhold information about their health.

Matheson was sixty-five years old when he first saw Dana. Matheson was an ordinary-looking man, with a balding head, wire spectacles, and a round, inconspicuous face. His erect posture and slightly distinguished appearance, however, gave him the commanding air of power. On his case history, taken July 13, 1922, the diagnosis reads "*Parkinsonism*," the classic chronic symptom of epidemic encephalitis. Dana included some personal information: Matheson was married with three children and had a brother who died of consumption—it is not known if that was the same brother who survived the shipwreck with him on their way to school in Scotland. Dana included correspondence in the file and Matheson's own history with this disease—including his frustration with the many doctors he visited, so little information, and the long list of medications. Dana, or his assistant, typed up the report on Matheson's physical examination—a masklike face and a monotone voice that had improved, as well as normal blood pressure, normal visceral reflexes, normal weight, healthy appetite, and clean tongue.

Matheson spent nearly a decade trying several different courses of treatment, including some very experimental ones, and he began consulting Tilney at the Neurological Institute. He had essentially offered himself up as a guinea pig for several new theories. Given Matheson's dogged determinism, his own personal interest in the mysterious disease, and his wealth, it is not surprising that epidemic encephalitis became Matheson's own biomedical research project. Matheson's case of epidemic encephalitis would be a turning point in the history of this disease and breathe new life into medicine's struggle to control it. He created the Matheson Commission, a personally funded team of physicians to work out of the Neurological Institute, researching encephalitis lethargica and developing a vaccine. Then Matheson began the search for a qualified staff and talented leader.

CHAPTER 19

Josephine B. Neal

Social change often happens gradually like a large swell moving soundlessly across the sea, but when the wave reaches land that quiet swell comes crashing against the shore. When modern life began changing at the turn of the century from a farming focus to an urban one, it set in motion the deepwater current.

The shift seemed simple enough at first: people were moving from farms into cities. Trains now brought farm foods into urban centers. Foods and canned goods could be stored in pantries for extended amounts of time. With the advent of the icebox, milk and meats could be kept cool. Washing machines, vacuums, and other appliances simplified household chores.

Those simple changes would have a significant impact on the role of women.

While their days used to be regimented—washing day, ironing day, cleaning day, canning day, mending day, baking day, and Sunday dinner days—their home lives were becoming more modern. It was estimated that women on farms spent nearly forty hours a week on cooking alone. Without the need for farmhands, fewer children were needed to help tend crops and watch after the younger children.

Likewise, with a vast number of childhood diseases and epidemics now under control, there was little need for a larger family to compensate for the inevitable loss of the more fragile lives. The women's movement, too, was playing a part. With the founding of the American Birth Control League in 1921, women were for the first time in history educated about how to control the number of children they had. They could opt for two instead of ten. Fewer children meant more money for education, and girls, who had been traditionally overlooked because funds went to the education of their brothers, were able to enter schools and colleges. Education gave women a taste of freedom, and consequently, the wave of social change came crashing onto the shoreline.

This was a time, after all, when women had to fight just to be considered something other than the property of their husbands. Most women did not have an income, did not own property, and had no voice in political elections. In some sense, the dramatic change in appearance became like a psychological rebellion. Victorian clothing concealed everything—long hair was worn in buns or hidden beneath hats. Corsets bound breasts, and Victorian lace, pleats, and collars camouflaged them. Long skirts and petticoats drew a heavy curtain over legs. A woman's physical appearance was considered her husband's as well. Given that symbolism, it was hard to miss the message that bobbed hair, lower necklines, sleeveless dresses, and short hemlines sent.

Suffragists were also pushing for women to have a vote in politics, or more accurately, be granted the vote *again*. Some women in colonial America did vote before laws were passed denying suffrage to women, immigrants, and "people of color." The idea that "all men were created equal and endowed by their Creator with certain unalienable Rights" seemed to have been misinterpreted. Lest that misinterpretation be blamed on politics alone, certain sociologists, historians, and religious denominations made sure to solidify those beliefs. A leading French social psychologist wrote that scientists who had studied the subject knew women's intelligence to be one of the most inferior forms of human evolution, closer to children and savages than to an adult, civilized man. In case anyone missed his subtle point, he added: "Without doubt there exist some distinguished women, very superior to the average man, but they are as exceptional as the birth of any monstrosity." And it had been famously said in the South: "A woman's name should appear in print but twice—when she marries and when she dies." The quote was still relevant enough in 1918 to appear in a book on social progress—or the lack thereof.

Social change and education brought women into professional spheres in greater numbers than ever before—spheres that were not necessarily welcoming. While the number of women in law and doctorate programs doubled and tripled, medicine claimed only 4 to 5 percent of total enrollment. The idea of "separate, but equal" medical schools had started decades earlier, allowing women to attain a degree at a women's medical school. But it was a double-edged sword that did not translate into egalitarian opportunity. Hospitals and medical practices were quick to condemn those same schools they required to be segregated. And even the accusation that women's medical schools were ill-qualified was false. One medical historian later pointed out, "Vassar, Bryn Mawr and Smith had been sending large numbers of well-prepared women to Johns Hopkins for over two decades." Regardless, the percentage of women in medical schools remained low, in large part due to the quota system, which allowed for only a small number of women, Jews, or African Americans per class.

Women, it was believed, simply didn't have the mind for science or medicine—in spite of the fact that Marie Curie had just become the first person to win the Nobel Prize *twice*.

★ ★ ★

It is significant then, that when Matheson sought a leader for his newly formed Matheson Commission, he chose a woman. Dr. Josephine B. Neal was an expert on encephalitis; even more importantly, she was both a neurologist and a bacteriologist.

Neal, born and raised in Maine, had originally been a schoolteacher. When she saved up enough money, she applied for medical school, hoping for a profession in "the healing art." She earned her degree at Cornell Medical School, one of the first schools to accept women students. Little is known about her years in medical school and immediately following graduation. She lived in New York, in Gramercy Park, most of her adult life; she never married. One of the few existing photos shows her to be serious, with small, round glasses and wearing her gray-black hair pulled into a bun. She looks more like a schoolmarm than a leading medical researcher. Neal was a prodigious record keeper, but a scientific one, which apparently left little time for personal documentation. What is known is that by 1926, Neal had been appointed head of the meningitis division of the New York City Department of Public Health, working under the famed Dr. William Park, head of the bacteriology labs.

In the male-dominated medical world, many women like Neal forced themselves upon the medical community by becoming an especially valuable commodity. One area in medicine where they excelled was public health. By the turn of the century, public health was losing its luster, and women physicians were drawn to the plight of uneducated mothers, working to lower the high infant mortality rates, and child advocacy.

Women then won a legal and political victory in passing laws that provided public health education and care for mothers and infants. Women health officials and physicians also aimed to reduce the shocking number of deaths during childbirth. Male physicians argued for medical doctors and hospitals in the delivery of children; public health advocates argued that two to three times as many women died in a hospital than under a midwife's care. The primary reason for this was unsanitary practices. In hospitals, a physician would examine diseased patients, then walk down the hall to deliver a baby—without ever washing his hands. Most women who died in childbirth died of infection known as "childbed fever." The medical community compromised by requiring medical education or licensing for all midwives. Likewise, more women physicians began entering the field of obstetrics, demanding better conditions for both mothers and infants in hospitals.

With a strong focus on educating immigrant populations, protecting maternal and child health, and preventative medicine, public health was becoming known as "the woman's branch" of the government. Soon, there were too few men left in the field.

Neal's work in public health afforded her an excellent view of epidemics firsthand. She followed the outbreaks, identified patterns among neighborhoods or people, studied the tissue samples, and organized all the information. Meningitis is akin to encephalitis in many ways: meningitis is a swelling of the membrane around the brain or spinal cord, while encephalitis is a swelling of the brain itself. Both can occur in bacterial form, viral form, or as the body's own reaction to something. In both, the swelling can lead to permanent brain damage. Interestingly, meningitis itself can cause encephalitis.

When epidemic encephalitis first appeared, it was often misdiagnosed as meningitis. The symptoms are almost identical—severe headache, a stiffness in the neck, sensitivity to light, fever, and a change in mental function. But autopsies and tissue samples showed early evidence of epidemic encephalitis and not meningitis. So, in a sense, Neal had been tracking encephalitis lethargica from the start.

Neal was also considered an expert on polio, a viral disease that causes swelling of the spinal cord, often resulting in paralysis. It is well-known that Jonas Salk tested the first polio vaccine in the early 1950s and introduced it publicly in 1955. What is not as well-known is that the laboratory for the health department of New York City was working on antipolio vaccines as early as the 1930s. The vaccine moved quickly into human trials and, true to her determined nature, Neal was one of the first people to take an injection in 1934. She fared well, but others did not. Severe allergic reactions occurred in some cases and even polio in others. These results were generally kept out of the media, and the vaccine campaign stopped until the 1950s.

It is easy to see then how Neal's expertise would cross several lines—public health and private practice, viral agents and bacterial ones, swelling involving any part of the brain or spinal cord, neurology and infectious disease studies, even vaccine development. There could be no better choice to head the Matheson Commission.

In a conservative wool dress and sensible shoes, Neal left her home in Gramercy Park for her office. Noise was everywhere. Trolley cars had stopped running in many of the streets in her neighborhood because they slowed car traffic. By then, the number of automobiles—New York City alone had more cars than all of Europe—caused a dense smoke cloud to settle over the city. So Neal took the Lexington el that rose out of the streets, buttressing the bright orange train cars that came thundering along the tracks, sprinkling pedestrians with fine, powdery black soot.

On street corners, cinders, wood ash, and coal dust were swept into large piles, near stacks of crates, wastepaper cans, and empty bottles waiting for the garbage auto to pick up, to deliver and dump in the wetlands outside the city. If Neal had looked out the smudged windows of the el through the downpour of soot below her, she would have seen a sidewalk awash in dusty hats like raindrops hitting pavement.

There was at once an intimacy and a disassociation from the crowds in Manhattan. People moved independently of one another at the same time and in the same direction. Neal walked the rest of the distance among the purposeful throngs of strangers. She undoubtedly blended in with the crowd; none could have guessed who this seemingly unremarkable woman was or where she was going. Neal turned onto 105th at the New York Academy of Medicine. It had been decided that Neal's office should be located there, where she had access to countless medical articles and information on public health. The academy boasted, and still does, a collection of medical literature dating as far back as 1700 BC. The great, Romanesque building was only two years old but had been endowed with a timeless, traditional interior. Its library had dark wooden floors and tall, beamed ceilings with large, arched windows that looked out onto Fifth Avenue and Central Park.

Neal entered the building, and the elevator operator pulled the cage door to take her up to her office, where Neal sat down and began to absorb the enormous task facing her. The chronic effects of epidemic encephalitis were now considered even more debilitating than polio. A *New York Times* article summed it up well: "Because it particularly attacks young people and either kills them or leaves them incapacitated for the rest of their lives in many cases, medical authorities regard epidemic encephalitis as one of the most imperative problems the science of medicine has to solve."

CHAPTER 20

Vaccine Trials

In the era since the creation of the Centers for Disease Control and Prevention in 1946, medical investigations have become large-scale and fascinating to the general public. The bacterium that causes Legionnaires' disease, the Marburg virus, the Ebola virus, and HIV conjure a palpable fear because they are relatively new to us, the diseases they cause can be lethal, and there is no vaccine. The word "outbreak" brings to mind secured labs, biosafety levels, and hazmat suits. And yet, it is essentially the same type of research conducted by medical investigators during the 1920s, '30s, and '40s. Polio, yellow fever, smallpox, measles, influenza, and syphilis were the great fears of that age, and the epidemiologists who studied them did so in unprotected clothing in a basic laboratory, often testing the vaccine or treatment on themselves.

Essentially, medical investigation requires that a number of things come together. It takes a benefactor, a team of talented scientists, and, of course, a disease outbreak. All three would normally move independently of one another, but, like the three hands of a clock, every so often they line up perfectly. In 1927, just that happened.

Matheson had been suffering for nearly a decade from the chronic aftereffects of his case of encephalitis lethargica. Like most of the patients, he could see the bleak future ahead of him, the slow decline into immobility. Every tremor or stunted movement was a reminder that he was beginning to lose control. For a man like Matheson, that was unacceptable. Matheson's original budget for his commission allowed for $10,000 per year, by today's standards around $120,000.

In addition to Neal, the Matheson Commission was composed of a number of New York neurologists, including Tilney, and specialists in public health, like Park. A major goal of the Matheson Commission was to bring together scientific research with public health. It was this intimate relationship between New York's public health and the neurological research that set New York apart from other cities, like Chicago and Philadelphia, struggling with the problem of epidemic encephalitis.

While Neal's office remained at the academy, the rest of the commission was headquartered in the new, state-of-the-art Neurological Institute, which itself was to be part of the new Columbia medical facility. Columbia's research and hospital facilities rose in great brick blocks out of the Washington Heights neighborhood. Through its windows, there was an unobstructed view of Manhattan stretching out before it, the Hudson, columns of smokestacks in neighboring New Jersey, and, just in the distance, the construction of the polished Washington Bridge reflecting silver light off the water.

The Matheson Commission took on this epidemiological endeavor to study sleeping sickness, but it also aimed to treat the victims themselves. That in and of itself was a unique approach to a medical investigation. In most cases, an epidemic disease and its relationship to people are the focal points of research. Patients either die of the disease or recover, and the work of the investigators is to figure out how the disease is spreading and create a solution or vaccine to stop it. The Matheson Commission faced a unique challenge. Victims of encephalitis lethargica who survived did not necessarily recover. Instead, they lived in medical purgatory, neither dead nor well. For the Matheson Commission, helping the patients as their brains slowly eroded was just as critical as creating a vaccine.

The first thing the commission needed to do was get a full understanding of this pandemic. Few diseases have had such a thorough and organized investigation in so short a time. And this was no easy disease to chronicle. Investigators did not have a definitive test to determine encephalitis lethargica, and the disease was often misdiagnosed. The outbreak occurred in two phases, with a first wave of initial acute cases, followed by a second wave of chronic cases. Complicating things further was the fact that acute cases could go completely unnoticed with mild symptoms, and only years later, when the chronic symptoms like Parkinsonism or mental impairment surfaced, did doctors realize the patient had suffered a case of the disease. Epidemic encephalitis had been confused with everything from influenza to polio, depression to schizophrenia. In order to get any sort of handle on this disease, the physicians first needed to determine exactly how many cases there had really been.

The New York–based group of scientists decided to catalog not only the thousands of cases in New York, but also cases of the disease abroad. Neal sailed for Europe in 1928 to visit the Lister Institute, Sheffield Medical School, and the National Research Council in England, then clinics at The Hague, the Pasteur Institute in Brussels, and the Pasteur Institute in Paris to learn their theories and treatment plans. She also planned to meet with Arthur J. Hall, who first recorded the outbreak in London. Meanwhile, Tilney recommended that there be a network of neurologists in this country, so that doctors in cities throughout the United States could notify the

Matheson Commission of cases and treatments they found effective. Another committee member recommended that a federal board be established in Washington to document any outbreaks of this disease in the country.

After the commission gathered all of this epidemiological material, they organized it into three separate Matheson reports, as well as a book authored by Neal in 1931. The Matheson Commission found that encephalitis lethargica had been reported all over the world: Ireland, England, France, Holland, Sweden, Austria, Algeria, Greece, China, the Philippines, Cuba, Brazil, and Canada, not to mention throughout the United States. The commission also compiled everything that had ever been printed about epidemic encephalitis—in all languages. They had articles from widespread places: from China to Chicago, Australia to Siberia. They also found at least seventy-five different types of treatments, testament to how little the medical world knew about dealing with this disease. Neal managed to compile around one thousand case studies for the Matheson Commission, and over 80 percent of those would be seen at the Neurological Institute.

With a firmer understanding of this pandemic's immediate history, the Matheson Commission moved on to its next and most important task: developing a vaccine and testing it on humans. With theories about this disease covering the full spectrum, it really came down to two questions. Was this disease *caused* by a particular bacterium or virus, or was the encephalitis the body's reaction to a typical infection elsewhere in the body? And second, why was encephalitis lethargica occurring in epidemic form? There were three fairly solid theories as to what the culprit was, which determined the paths the vaccines would take.

One group of researchers believed epidemic encephalitis was

caused by an unknown, filterable virus, much like the one that caused polio. Find that virus, and they could create an attenuated vaccine—one with a harmlessly low level of the virus, but enough to prompt the body to produce antibodies. With the electron microscope a few years away from completion, some microbes themselves were not even visible. Researchers passed blood samples through filters and took it on faith that something was there, even if they couldn't see it, calling these invisible invaders "agents" that caused an infectious disease.

Another group focused on the search for a bacterium, primarily a streptococcus, that could be the culprit. In this case, it was believed a bacterium was causing localized infections in the mouth. Culture the bacterium, and a vaccine could be made out of bacterial cells. The second vaccine, known as the "Rosenow vaccine," was named for Dr. E. C. Rosenow of the Mayo Foundation, who not only treated patients with his form of vaccine, but also removed their teeth to stop the infection's point of origin.

The last theory was that epidemic encephalitis was linked to the herpes virus. The vaccine was created by Dr. Frederick Parker Gay, who developed "A and B vaccines," which would test a control group of patients against a group given a vaccine made from hyperimmune rabbit brain tissue. In the case of the A and B vaccines, the vaccine would let the animal build the antibodies and lend them to the human.

Vaccine development is notoriously complicated and fickle. For one thing, scientists are often dealing with a "living" organism. Too little of the agent in a vaccine is useless; too much can infect the patient with the very disease it's meant to prevent. "Dead" viral or bacterial cells often fail to push the body's immune system into action. And, of course, researchers always take a risk when injecting foreign material into the human body. It might elicit an immune response, just not the one they were aiming for. If the body rejects the foreign substance, it might respond with any number of "post-vaccinal" complications—most notably, encephalitis.

The vaccine trials were quickly under way, with tests on patients at the Neurological Institute, as well as a group of children at Kings Park State Hospital. No armchair benefactor, William Matheson himself was a test subject. His personal physician since 1928 had been Dr. Rosenow, and Matheson believed he was receiving great benefit from the Rosenow vaccine and had even undergone Rosenow's "treatment" to have many of his teeth removed. Matheson may have been one reason Rosenow's vaccine was still in the race. Rosenow himself was regarded as "unbalanced" and "paranoid" by some of his colleagues. Most researchers showed little faith in his streptococcus bacterium theory, not to mention his treatment for it. Virology was new and exciting; it was the future of medicine in the 1920s. Bacteriology, in its heyday during the late 1800s, was the distant past.

Rosenow also came under attack from Neal, who did not much believe in his theory. Infighting was inevitable among so many talented scientists who held fame and a new vaccine within their grasp. About this time, Neal came under attack from the rest of the Matheson Commission. It's hard to know how much was valid and how much was driven by the fact that a woman was at the helm. Dr. Kenton Kroker, a medical historian writing about the commission, said that "Neal's position as a woman made her somewhat of an outsider within the overwhelmingly male circles of medical administration, practice, and research—yet she emerged as the single most important member of the Matheson Commission."

A flurry of letters back and forth between members of the commission and Matheson either condemned Neal or requested outright that she be replaced. The letters carry the male camaraderie one would expect from that time. Some are addressed to "My Dear Skipper" and peppered with references to the week they spent yachting together, as well as greetings from their wives. But when the subject of Neal is addressed, the tone changes.

In what must have been a humiliating turn of events, Neal was asked to leave a Matheson Commission monthly meeting so they could discuss *her.* The member writing a letter to his dear Skipper Matheson described it: "I then brought up the question as to whether her personal delinquencies as to irritating people were sufficient to outweigh her other qualifications for the work and whether to continue with her or endeavor to find someone to take her place." After much discourse on the subject, it was decided unanimously that "aside from her personality, she was the best qualified person for the work. . . ." Neal's "lack of tact and loquacity" would be brought to her attention by the commission, so that she might "mend her ways."

While Neal irritated a number of the men she worked with, others were surprised by the reaction of their male colleagues and enjoyed working with Neal, especially Tilney and Park. In fact, Park had hired a woman for his bacteriology lab when New York's health department was at its strongest. Her name was Dr. Anna Williams, and she was one of the best bacteriologists in the country, with a keen ability to find anything beneath the microscope. Still, many male physicians in the city ridiculed Park's work with both Williams and Neal, referring to the lab as Park's "harem."

It was decided that two cases of the flu and no vacation time in a year had probably played a role in Neal's personal demeanor. Either way, the one thing they all agreed on was that she was too talented and qualified in this field to let her go. Tilney even recommended that she step up her involvement, maintaining her ties with the health department and having greater involvement in the Neurological Institute, as it would "enable her to be in touch with different foci of encephalitis throughout the City and also make it possible for her to have official contact in follow-up. . . ."

★ ★ ★

During the vaccine trials, the Morgan Ward of the Neurological Institute served as the center-ground for the commission. Neal hired a full-time neuropsychiatrist, three nurses, and a secretary for the Matheson Commission alone. Patients would come into the clinic at the institute to be studied, treated, and given vaccines, but the physicians would also visit the patients' homes, follow up on how they were feeling, and keep detailed notes. The vaccine trials were certainly "on the encouraging side." The commission rarely got to see acute cases of the disease—after all, the epidemic was considered over by then. About one-tenth of the patients in the trials were in the acute stage, but their death rate was cut significantly when the vaccines were used.

Vaccines were not the only treatments for these patients either. The doctors had been experimenting with vitamin B injections and "the Bulgarian treatment," which had showed some success abroad and used the roots of the belladonna plant grown in Bulgaria. The treatments gave relief to patients at times, although it's not known if it was a psychosomatic response or a real one.

In chronic cases, the vaccines were showing mixed results. Still, the need for a vaccine was becoming desperate. The chronic symptoms of epidemic encephalitis had been appearing in greater numbers, just when the doctors were trying to get a handle on the epidemic itself. Cases like Rosie's, where there was a severe personality change, were more common among children, but by far the most common chronic effect of encephalitis lethargica was Parkinson's disease. The decline of these patients was inevitable and progressive. Without some type of vaccine or treatment, these patients—thousands of them—would slowly freeze into an immobile state. Since they were unable to communicate, care for themselves, feed themselves, or even walk, the only places they could go would be mental institutions. And so by 1930, there were waiting lists for the vaccine.

Because the chronic nature of the disease presented an unusual challenge for the researchers, one member of the commission explained, "The only way in which real clinical knowledge of this baffling disease can be obtained is through the study of a large number of cases over a long period of time."

The Matheson Commission would not be given that chance.

It must have been a mild day in Miami on May 15, 1930, when William Matheson, aboard his yacht, the *Seaforth*, returned to Miami after a trip to the Bahamas. About an hour from port, Matheson suffered a massive heart attack and died.

For the commission Matheson had created, his death was not only a personal loss, but also a financial one at a critical point in the vaccine studies and disease survey. His death also came at a dismal time in the city's economy, in the very thick of the Great Depression. In his will, Matheson left shares of a corn refining company to the Matheson Commission, but the stock market crash had diminished their value. It would not be enough to fund the large project for much longer.

Dr. Neal's title had been changed from Director of Research to Executive Secretary because it was felt that the whole group conducted research, and Neal was focusing on the compilation of material. After Matheson's death, however, Neal behaved not like a member of the team, but its leader. As funds dried up and vaccine production slowed, she went to work raising money. First, she voluntarily reduced her own salary by 40 percent. She gathered contributions from friends, raffled off an automobile, and managed to raise about $5,000 on her own. The commission gave the money to a Wall Street investor, which turned out badly, reducing the sum to just over $3,000. There was barely enough money to finish the book Neal had assembled.

Neal's staff was reduced to just one nursing assistant. Still, there were enough vaccines and plenty of patients to keep the study open on a much smaller scale for another decade. The Morgan Ward of the Neurological Institute was becoming like a charity clinic. The sudden appearance of St. Louis encephalitis in the 1930s also furthered the need for research. Although St. Louis encephalitis is a virus spread by a mosquito, much like West Nile, it was nonetheless an encephalitis and fell within the realm of their research.

For the Matheson Commission, their work wasn't just about the race for a vaccine, or even the opportunity to make New York's Neurological Institute famous for its encephalitis research. It was also about the pitiful and hopeless condition of its patients, patients with whom the doctors had developed long-term and friendly relationships. One of the doctors on the committee wrote in a report, "The importance of a solution to the problem of encephalitis cannot be underestimated. The existence in the United States of more than fifty thousand patients . . . suffering from varied and disabling symptoms of the chronic stage of the disease constitutes a humanitarian as well as a real public health problem." He went on to describe it as a "truly tragic condition," with disabilities "infinitely greater" than those of polio survivors. The physical handicaps of the adults and the mental defects of the children who survived epidemic encephalitis became an enormous strain on their families. The clinic, it was believed, gave these patients a sense of hope in what would otherwise be a bleak future.

Neal was able to keep the Matheson clinic in operation throughout the 1930s, in spite of the economic depression. She continually made the trip to her office at the New York Academy of Medicine. On other days, she took the Second Avenue el the length of Manhattan toward the Bronx and any number of hospitals along the East River. Glimpses of the river flashed at each cross street in frames like a motion picture. In the winter, when dusk came early, the view was dark until the bright bulb of the lighthouse off the tip of Welfare Island ignited like a false moon over the river. It must have cast a lonely light on the institutions lining the island (which people of the previous generation still referred to as Blackwell's Island), especially the eight-sided dome of the lunatic asylum that sat closest to the lighthouse. The sliver of land parallel to the east side of Manhattan, yet still distinctly separated by ferries, the shadow of the Queensboro Bridge, and rough water, must have seemed an obvious symbol of isolation. There was a prison that had recently moved to Rikers Island, a hospital for the chronically ill, the lunatic asylum, a smallpox hospital, and now an influx of the downtrodden during the Depression—it was a solution for those who no longer had a place on the larger, neighboring island of Manhattan. If progress and change were to succeed in the 1930s, society had to move forward, not stall under the weight of challenges. Despite all her efforts, it was a lesson Neal soon learned.

Slowly, the neurologists involved in the encephalitis lethargica research project began to turn their attention toward other diseases, even seeing vaccine production as getting in the way of other research. Worse, New York and its Neurological Institute were no longer considered the center of neurological study. Both Boston and Philadelphia now boasted their own impressive neurological centers. As a result, the number of patients they were able to accept into the vaccine trials at the Neurological Institute was greatly reduced. One of the only doctors Neal still had on her side at this point was Tilney, who remained interested in the clinical trials and possible treatments for his own patients.

As fate would have it, Constantin von Economo also died suddenly in 1931. By the time of his death, he had been nominated for the Nobel Prize in Medicine three times for his research on encephalitis lethargica. Undoubtedly, von Economo regretted not living long enough to see any solutions to this cruel disease that would continue to bear his name for decades to come.

* * *

Another blow to the Matheson Commission during the 1930s was the sudden popularity of a new technique known as psychosurgery—the most famous of which would be known as the frontal lobotomy. The first generation of neurosurgeons had been born out of this time period. Still a brand-new breed, neurological surgeons of the 1920s and '30s had all been trained as general surgeons—there were no practicing neurosurgeons to teach them.

When neurosurgeons operated, they carried their own instruments to the various hospitals. The surgeons knew enough to wash their hands for at least ten minutes and to soak their scalpels in iodine and alcohol, and they'd learned, when drilling burr holes in the skull, to save the bone dust to fill in the holes afterward. Even if the neurosurgeons took precautions, however, the hospitals they worked in were often unsanitary. At Bellevue, where New York's indigent and poor flocked, surgeons sometimes kicked away the cats and rats beneath the operating table while they worked. One patient, during a brain surgery in which she was still partially awake, asked, "Do you-all mind if I have just a puff on a cigarette? I brought 'em along just in case." The surgeon replied: "Go right ahead, honey."

A nurse also told the story of how they humbled the new neurosurgeons at Bellevue. With windows facing west for better sunlight, operating rooms could become stifling. The attending nurses took salt tablets before the surgery, giving no warning to the eager young surgeon. An hour or so under that kind of heat, and the surgeon would faint with his amused staff still standing around him.

With little definitive success among the vaccine trials, psychosurgery and its building popularity became another option for patients with chronic epidemic encephalitis. It would seem a lucky stroke of fate for a patient named Sylvia.

S. E. Jelliffe, a friend of Sigmund Freud's and Carl Jung's, became one of New York's finest psychoanalysts. Jelliffe believed in keeping neurology and psychiatry a united force in brain study. His nonconformity kept him on the fringe of New York neurology through much of his career. *New York Psychoanalytic Society and Institute Archive*

When less than 5 percent of medical students were women, Josephine B. Neal became a bacteriologist, professor of neurology, public health researcher, and leader of the Matheson Commission. *New York Academy of Medicine Library*

Frederick Tilney, a gifted student and visionary, became a leader in American neurology. Tilney's book *The Brain: From Ape to Man*, was considered "one of the finest pieces of evolutionary writing since Darwin." *Archives & Special Collections, Columbia University Health Sciences Library*

An aviator, an aristocrat, and a pioneer in neurology, von Economo first recognized the disparate set of symptoms as part of a single disease. He named it *encephalitis lethargica*—literally, the swelling of the brain that makes you sleepy.

LEFT: Jelliffe both lived and treated his patients in a town house on Fifty-sixth Street between Fifth and Sixth avenues. Greenwich Village muse Mabel Dodge, actors John and Lionel Barrymore, members of the Algonquin Round Table, even the mistress of New York City's mayor visited Jelliffe here. *Author's Collection*

RIGHT: In 1926, this Romanesque style building became the home of the New York Academy of Medicine and a meeting place for the New York Neurological Society. Even today, the building remains relatively unchanged and houses an impressive collection of medical texts dating as far back as 1600 BC.
Author's Collection

The Neurological Institute, part of Columbia University's Medical Center in Washington Heights, was built to be the world's greatest center for brain study. The new building opened in 1929, offering groundbreaking psychiatric care, neurological treatment and surgery. *National Library of Medicine*

New York Neurological Society

BALLOT FOR 1927

President
George H. Kirby

Vice-President
(Vote for one)

~~Joseph Byrne~~ Louis Casamajor

Secretary and Treasurer
Charles E. Atwood

Councillors
(Vote for five)

~~L. S. Aronson~~	~~B. Sachs~~
S. P. Goodhart	~~M. A. Starr~~
J. R. Hunt	F. Tilney
S. E. Jelliffe	~~W. Timme~~
~~M. Osnato~~	E. G. Zabriskie

A voting ballot for officers of the New York Neurological Society.
New York Academy of Medicine Library

New York Neurological Society

Meeting of May 6th, 1919

Stated Meeting

PROGRAM

I. ***Presentation of Cases.***

If possible, several cases of "Encephalitis Lethargica" will be shown by members of the Committee on Encephalitis Lethargica.

II. ***Papers.***

1. Clinical experiences with Epidemic Central or Basilar Encephalitis.
   ("Encephalitis Lethargica")
   Dr. Bernard Sachs.

2. Committee Report on Encephalitis Lethargica.
   Dr. Isador Abrahamson, *Chairman.*

3. Observations on Gunshot Wounds of the Head.
   (With lantern slides)
   Major K. Winfield Ney, U. S. A.

Discussion by Drs. Charles A. Elsberg.
Alfred S. Taylor.
Harold Neuhof.

Epidemic encephalitis, as it was called in the United States, appeared many times in the society's programs.

New York Academy of Medicine Library

As city buildings reached new heights during the 1920s and '30s, the streets themselves became valleys to a mountain range of concrete, steel, and glass. *Library of Congress*

New York City rose from the depths of the Great Depression to host an exposition dedicated to the future. One hundred fifty years after the inauguration of George Washington his statue stands before the emblems of the fair: the Trylon and Perisphere. *Library of Congress*

A moving walkway carries people through the Trylon and Perisphere. Visitors wore buttons reading "I Have Seen the Future," and were introduced to technological breakthroughs like the calculator, electric dishwasher, fax machine, and television. More than 40 million people visited the 1939–40 World's Fair. *Library of Congress*

Window light ignites the Empire State Building. New York's sky-scrapers, the tallest in the world, not only boasted magnificent views of the city, but also became part of the urban panorama.
Library of Congress

A New York City letter carrier during the 1918 influenza pandemic.
Archives and Records Administration (Image 165-WW-269-B15)

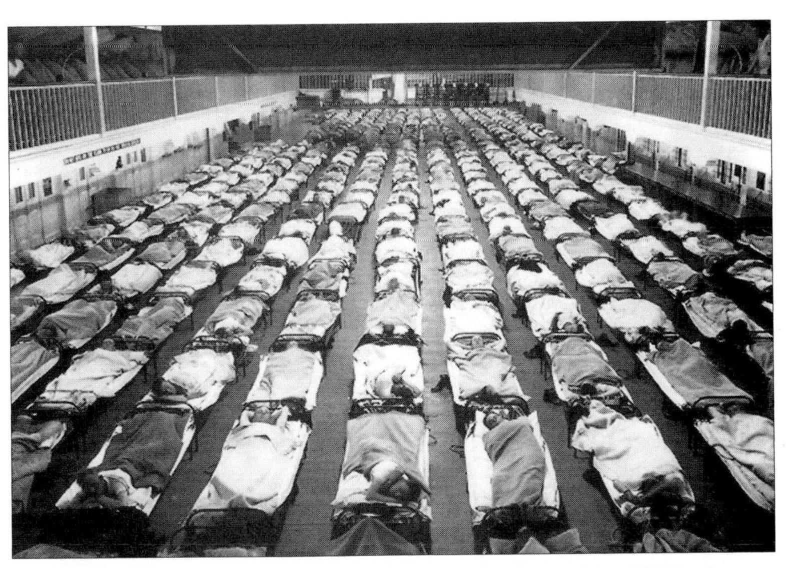

Soldiers sleep in a Naval Training Station during World War I. Their beds alternate head to foot to prevent further spread of diseases like pandemic flu. *U.S. Naval Historical Center*

Kings Park, a farm asylum on Long Island, hoped to be just that—an asylum from the hectic, nerve-wracking city life. Patients moved freely among the community and helped maintain the self-sustaining farm. *Kings Park Heritage Museum, Pete Hildenbrand Collection*

The Children's Unit of Kings Park was opened primarily to handle the children suffering long-term effects of encephalitis lethargica. The children were offered school lessons, baseball games, movie nights, dances and birthday parties. *Kings Park Heritage Museum, Pete Hildenbrand Collection*

An abandoned cottage still stands on the asylum grounds. The children's cottages were very similar in design and make to this now-vacant one.
Author's Collection

The Children's Unit once stood just beyond this solitary tree trunk. The children's cottages eventually were razed to make way for newer, larger buildings needed to house the growing number of patients. *Author's Collection*

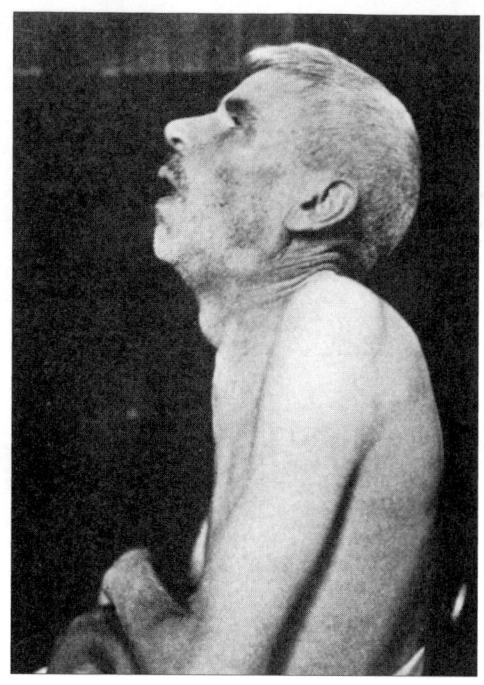

A patient of Dr. Frederick Tilney's endures an oculogyric crisis, a type of physical spasm that was a typical symptom of post-encephalitic Parkinsonism. One patient described it as "Jesusly painful," and it could last for hours at a time. *New York Academy of Medicine Library*

New York City went from being one of the world's filthiest and most disease-ridden cities to boasting one of the greatest public health departments in history. *New York City Municipal Archives, Department of Records*

New York's Public Health Committee once estimated that the city's regular tons of ashes, garbage, dead animals, street sweepings, and rubbish piled together would be higher than the Woolworth Building. *Library of Congress*

Women in particular took an interest in public health, bringing maternal and infant care to the attention of city leaders. *New York City Municipal Archives, Department of Records*

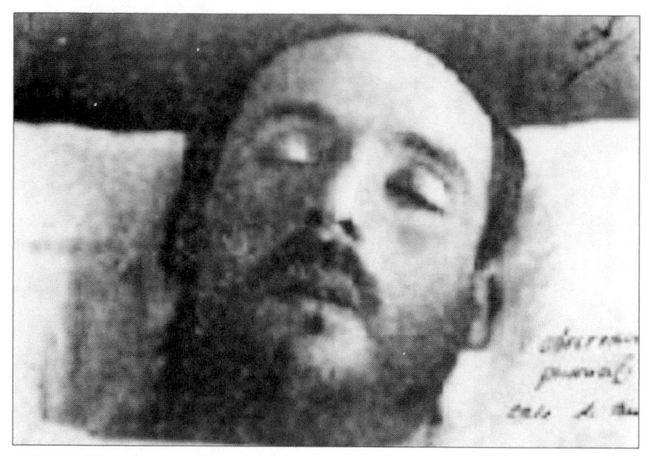

Patients suffering from encephalitis lethargica were not comatose, but asleep. They could be roused and awakened for short periods of time, but eventually fell back into a deep slumber that could last weeks or months.

The original tale of *Sleeping Beauty*, first published in 1697, may have been inspired by early epidemics of sleeping sickness in Europe. *The Rose Bower, part of the Legend of Briar Rose series by artist Sir Edward Burne-Jones.*

CHAPTER 21

Sylvia

September 24, 1942
Miss Sylvia Williams
Hall-Brooke Sanitarium
Greens Farms, Conn.

My dear Sylvia:

Your sister has just written to me about the holes in your head. There is nothing extraordinary about this because in doing an operative flap, four small trephine or burr holes are made. Between these a tiny wire saw is inserted and the bone is cut smoothly with a beveled edge. After the operation this so-called flap is put back in position and the beveled edge grows together. The burr holes remain, though after several years they may fill in with new bone. However, there is nothing to worry about since they are quite normal and no more of them will appear. So you need not worry about that.

Your sister tells me that you are in a chair and are feeding yourself. I am delighted to hear about this. I urge that you do this and do everything for yourself that you possibly can for the more you do, the more you will be able to do.

Please have one of your nurses write to me, or if you can feed yourself, you can probably hold a pencil and both Doctor Cramer and I would like a note from you written by you, no matter how short.

With very kindest regards from both of us.

Sincerely,
Kate Constable, MD

Sylvia awoke to the sound of quiet voices and the percussion of machine gun fire in the distant, war-torn fields of France. She felt weak and feverish, and her chest weighed heavily against her body. It was 1918, and she knew it was a case of the flu she had seen spreading through the hospitals. It was strange to be in the hospital tent where she usually deposited patients, and odd to be treated by the same Red Cross nurses she worked with. It was usually Sylvia who drove the ambulance into the hospital station, and after a few more days of rest, she would return to work, fully recovered from the flu. She would return to wearing the muddy boots, the heavy belted coat with the white armband and Red Cross symbol, and to pinning her hair beneath the metal helmet. She would return to the boxy ambulance, marked with a cross, which was welded onto the frame of an automobile and cranked to start—headlights like rounded eyes, a front grate, and spoked wheels. Serving in France, she drove mostly ambulance autos, but in many places, the Red Cross was still using horse-drawn ambulances.

For a woman to be an ambulance driver in the war, she must have had a strong commitment to the cause. When it was decided more men were needed on the front lines, nearly five thousand women volunteered as drivers. They came close to the trenches, with shrapnel flying, and they gave first aid to patients before the ride to the hospital. Along the war-beaten roads there were crooked, headless tree trunks and gaping holes where German shells hit the soil with tufts of cotton-white smoke billowing above. And then the driver made the trip along ragged roads to the hospital, listening to the patients scream or moan each time the ambulance hit a hole or bump. Maintenance was also up to the driver, so women worked as roadside mechanics as well.

In fact, the involvement of women in the war helped push President Woodrow Wilson to finally support the suffragist cause—how could the thirty thousand women who enlisted in the U.S. Navy, Marines, Coast Guard, and nursing corps (military opportunities that were remanded once the war ended) be asked to defend their country without having a vote in its endeavors?

By coincidence or by fate, Wilson would have his own case of the flu in France, while negotiating the terms of peace after the war. Originally, Wilson's attitude toward Germany had been one of leniency. He disagreed with the harsh terms France and Britain proposed in the treaty, arguing that stripping Germany bare would be like building peace upon quicksand. He warned that Germany would emerge bitter and broken, and that "only a peace between equals can last." Wilson was so angry, he threatened several times to leave Paris altogether. But when Wilson was taken to his bed with what seemed to be a severe case of the flu, aides noticed that his personality changed overnight. One reportedly said that "something queer was happening in his mind." He went on to say, "One thing was certain: he was never the same after this little spell of sickness." Wilson grew increasingly paranoid and obsessive, also developing a facial tic. Biographers and physicians have examined his papers and debated the cause—whether influenza, a stroke, or encephalitis lethargica. There is no real way to know what caused the lasting damage to his brain. What was obvious, however, was his change in temperament and thinking. When he returned to negotiations, his views were reversed. He agreed to the harshest treatment of Germany, and his original prophesy of peace settling on quicksand proved to be true. Wilson's mental decline and physical handicaps were kept a secret from the public for nearly a year and a half and would later lead to changes in the disability acts surrounding the presidency.

It's interesting to note that Wilson's potential bout with encephalitis lethargica just after World War I and Adolf Hitler's possible case of the disease during the war would each land Europe on the path to World War II—Wilson for agreeing to the harsh Versailles Treaty, Hitler for rising out of the treaty's inherent dysfunction. Even more telling is how pervasive the fear of encephalitis lethargica was at that time. The disease was still so broadly defined and inconsistently diagnosed that this "forgotten epidemic" was becoming a catchall for erratic and damaged brain function.

In the winter months of early 1918, when Sylvia had influenza, hospital tents in the French countryside were filled with flu patients. Soon, the hospitals in and around the cities were as well—especially when large battles were expected, and the field hospitals had to be evacuated to make room for the wounded. In these wards already overcrowded with war wounds, especially with mustard gas injuries like swollen, disfigured faces and blindness, the flu cases were almost as frightening. Some people died within hours, literally drowning from the fluid in their lungs. The effect was so dramatic and so quick that it starved the body of oxygen, and flu patients grew cyanotic, with their toes or fingers or lips growing blue-black. Everything possible was done to keep the illness from spreading to other patients. The wards were so full, sometimes the only places the nurses or drivers could find to sit and rest was on the empty coffins.

Sylvia recovered from her case of the flu, continued her work as an ambulance driver, and, just as the Red Cross was withdrawing its workers in 1920, came down with another case. Her departure was delayed, but she recovered fully. She remained healthy and after the war spent time on the French Riviera. After that last bout with the flu, Sylvia described herself as "never ill a day." It was a full decade before that would change.

Sylvia was living in the United States again, both in an apartment in New York and at a family home in the Connecticut countryside, a sprawling, two-story white clapboard house with dark shutters and a winding porch. One day, while she was playing tennis, she suddenly threw her arm out of joint. She saw her family doctor, and after a few weeks, her arm was completely recovered—until it began to shake. A friend, who noticed Sylvia's arm was "out of pluck," recommended she wear it in a sling, which she did for the following year. The trembling in Sylvia's arm continued and would come and go depending on how she used her arm. She complained of some stiffness as well. By the time she visited the Neurological Institute in 1934, she was also walking with a limp.

Originally, Sylvia was seen in Dr. Tilney's office, with her medical record typed on his letterhead, but she later was examined by a doctor named Kate Constable. Given Sylvia's age, fifty-three, which was still considered young for the traditional onset of Parkinson's disease, and her history of the flu during the war, the diagnosis became chronic epidemic encephalitis. Remarks in her case history included, "This patient has a progressive, degenerative disease, the prognosis of which is very poor." Sylvia was showing the classic chronic symptom of epidemic encephalitis: Parkinsonism. Photographs of patients with advanced stages of this condition show them hunched over, the spine protruding awkwardly from the back, hands hanging well below the knees, fingers curled, like the evolution of Darwin's early man, but in reverse—a step backward from the upright, normal, healthy Homo sapiens. Documentary footage of the patients depicts them in much the same way, with arms frozen in front of them or knees permanently bent or bodies slumped forward in a wheelchair. They have the same masklike expression and smooth, waxy skin that robs them of their individuality and disguises them in uniformity. They look like bodies that have been stripped of their minds.

The footage is a sad foil against the advances in motion pictures during that time. Movies captured the movement of people; these films captured their terrible immobility. These statuesque victims of sleeping sickness are even thought to have been part of the inspiration for an entire genre of horror films. They began in German films with *The Cabinet of Dr. Caligari*, a film about a doctor—the head of an insane asylum—who uses a sleeping, entranced patient to murder people. America soon followed with a Bela Lugosi film whose tagline read: "She was not alive, nor dead." Another film told the story of a woman who caught a sleeping sickness and was the victim of shock therapy. These "zombie" films may well have been inspired by relics of the sleeping sickness epidemic. They showed in moving black-and-white the fear and dread of an illness that could cause deep sleep and permanent disability, a disease that left a patient neither dead nor fully alive.

Over the next six years, Sylvia would receive treatment in the encephalitis ward of the Neurological Institute fifty-one times. She sat in the soft pastel rooms where medicine cabinets held aspirin tablets, talcum powder, iodine bottles, boxes of cotton, and tins of Johnson & Johnson Band-Aids—and yet nowhere in those cabinets, among all those glass vials or cork-stopped bottles or tins, was there any medication to relieve her disabilities. A number of vaccines were tried on Sylvia, as well as bromide, vitamin B injections, quinine, and belladonna, which she detested because it tasted like prunes, but which she also felt was doing her much good. She even tried cobra venom after her sister read it could help with the aftereffects of sleeping sickness.

By then, the Great Depression was in full force. Most days, Sylvia rode in a taxi, with her dog in her lap, to the Neurological Institute for her regular treatments. The drive was long—autos rarely made it above fourteen miles per hour—and she must have watched with envy the women who were still able to walk and move freely on the sidewalks. Nonetheless, the somber mood on New York streets was evident. Women wore plain cotton ready-to-wear dresses, much less flamboyant than the styles of the previous decade. Rather than hand-sewn buttons, they had the more practical slide fastener—later known as the zipper. Shoes had sensible rounded toes and thick heels. An endless stream of black, gray, and brown clothes swept by the window of the taxi. Men, like the women, also dressed practically. They no longer changed suits during the day—it was becoming too costly—so men owned fewer suits or abandoned the suit altogether for jackets in blazing colors, or "blazers." On the sidewalk, they stood out like colorful leaves dropping in a flow of muddy water.

As Sylvia made her trip up to Washington Heights from her apartment on Park Avenue, she also passed the lunch counters and soup kitchens, storefronts boarded up and vacated, more beggars on the streets, and children asking for food. Street corners were bright with the reds and greens, pinks and golds of the apple carts thousands of New Yorkers now relied on to make extra change.

Surely, Sylvia herself began to wonder when the money would run out and how she would make ends meet. How much longer would the vaccine trials last, and if they proved ineffective, who would take care of Sylvia, as her condition grew worse and even more hopeless? Sylvia was seeking treatment on Mondays in the free clinic at the institute since she could no longer afford to pay her bills. Money was rarely out of her thoughts; she even named her dog Budget. Sylvia had never married, she had lost her job due to her disabilities, and she received only a small sum annually from the Red Cross. The rest of the burden fell to her sister and her sister's husband. This was not easy in the midst of the Great Depression. Sylvia's sister was already caring for their elderly father, as well as two adult sons unable to find work to support their own families. Her family had been selling property, their boat, giving up club memberships. They could afford only half of what was needed yearly for Sylvia. She had hoped to get a job herself, but jobs were scarce, and her handicap was becoming more apparent.

Sylvia's vaccine schedules were intense—weekly, then daily, with the measurements increasing slightly with each dose. A course could begin with 2 cc and end with over 20 cc. Sylvia kept accurate records, and indeed seems to have been prolific in her correspondence, even sending Christmas cards to her doctors that showcased her humor. The cards were from Sylvia as well as her aptly named dogs and cats—"Merry Christmas from Budget, The Vitamins, B1 + D, Calcium, Phosphorus + Sylvia Williams." She frequently sent telegrams asking questions of both Dr. Constable and Dr. Neal, who were overseeing her case during the vaccine trials. She was instructed to avoid coffee, spices, alcohol, and tobacco, but the last two were rescinded because Sylvia liked to smoke and enjoyed mint juleps in the summertime. The doctors had originally advised her to stop because those were the recommendations from Europe, but they were not necessarily adhered to in New York.

Sylvia rarely dealt directly with Tilney. For one thing, Tilney's health was beginning to trouble him by 1935. His position at the Neurological Institute was demanding, and he walked slowly, still limping from the stroke he suffered a decade before. But Sylvia did send him thank-you notes for her treatment. She also continued to write to Neal, inviting both her and Constable to visit her country home.

On some days, Sylvia went to the clinic for physical therapy, which included massage for her legs, in addition to her regular treatments. On other days, Constable would make house calls to Sylvia's apartment to give her injections. The first year went smoothly, although occasionally Sylvia's emotions got the better of her. She wrote an apology note to her doctor for letting her feelings get out of control at the clinic one day: "Up until then I had believed I was through with such reactions when problems presented themselves and I do believe my nerves are in a much better state of control than when I first went to the hospital. I never used to be this way at all. . . . " Another of Neal's patients at the clinic expressed similar views, describing herself as having a lack of charm, poise, tact, and diplomacy. "I am jealous, selfish and greedy," she said to Neal. "I use aggressive measures as defense mechanisms. I am become mentally subnormal and morally delinquent."

In general, Sylvia's case was typical of the disease in adults. It began with the Parkinson's symptoms—a tremor in the hands or the classic "pill rolling" gesture that looks like someone is rapidly rolling a small pill between the thumb and first finger. Sometimes muscles would freeze, and the patient would stop midstep. For patients with chronic encephalitis lethargica, the disease progressed from there, and it would for Sylvia as well.

Sylvia's letters chronicle her slow, steady physical decline—from a relatively normal life with a mild disability to one in which she was losing control. She staggered "like a drunk" from wall to wall when she tried to walk. But she was still typing her own letters and signing her own shaky signature. It's obvious from the number of them and the personal nature of them that Sylvia enjoyed writing letters. It was a way in which she could still stay connected as her disease grew more isolating. By 1940, she could no longer leave the house. On some days, she could not leave the bed. Her handwriting gave way. Soon, she would lose the ability even to take care of herself or feed herself.

Sylvia was losing weight, writing to her doctor that with a close haircut, she could look a lot like Gandhi. She described her legs as having no "starch," with weak, loose muscles that would manage to freeze up and partially paralyze her. "Too bad," she wrote, "one can't get spare parts."

The vaccines seemed to help Sylvia for a while. Or the apparent improvement may merely have been the course of the disease, which could progress at a slow but agonizing pace, leaving the patient hopeful one day and in despair the next. Sylvia let her apartment in the city go and spent most of her time confined to her country home not far from Danbury, Connecticut, receiving her vaccines and medications through the mail. Her gait had become so distorted that she had to bend way over at the waist in order to balance herself as she walked. And, unable to write, Sylvia now dictated letters to her nurse.

By the first days of 1940, she needed full-time care, and was also proving to be a handful for any caregivers. Sylvia's case became what so many of the chronic epidemic encephalitis cases did—a constant struggle for family to find a suitable place as the patient grew worse and more unmanageable. There was a steady stream of correspondence between Sylvia's family members and her doctors, most notably discussing their financial situation and the best place for Sylvia to receive care. At the time, small private "homes" existed for minor care of patients, but they could be costly. The vast majority of the homes for incurables were in awful shape—especially the affordable ones. Sylvia spent time at one of the more upscale homes, but described it as the closest thing to incarceration that she'd ever experienced—especially the strict "lights out" at 9:00 P.M.

By the following winter, Sylvia was back at her home in Connecticut, but it was her last winter there. She planned to stay through the pleasant summer months, but was already aware that by autumn, she would need a nursing home. She was bedridden now, unable to use her arms or her legs. Her family found for her what seemed like a perfect situation—a private home in West Haven, facing Long Island Sound. The woman had a state license to use her house as a nursing home with a maximum of six patients. In the beginning, Sylvia was the only patient. It wasn't long, however, before Sylvia's family received complaints from the house's operator that Sylvia demanded all the nurse's time, calling her four times in the night for various complaints or to ask to rearrange her covers an inch or two.

Her family realized Sylvia would not make it long in the private home, perhaps only until spring, before they might have to send her to some dreadful public institution. Out of desperation, they contacted Dr. Constable and asked for an honest assessment after her next visit to the Neurological Institute. If it was likely she would live only a few more years, they could afford to put Sylvia in a nicer, private institution; if not, they had to look at the most economical situation, which would be a long life in a "ghastly place." Dr. Constable wrote back, in December 1941, that Sylvia seemed in very good health, aside from her chronic epidemic encephalitis: "I think that one has to look forward to a fairly long period of life for her, unless, of course, something unforeseen happens."

Dr. Constable went on to mention that Sylvia questioned her closely for any new treatments or surgeries for people in her condition. She had heard of an experimental surgery in which the "pre-motor area of the brain is excised." In about 50 percent of the operations, some of the physical symptoms of chronic epidemic encephalitis, primarily the Parkinson's symptoms, could be alleviated. Dr. Constable told Sylvia she would have to gain some weight and get her blood count up before they could consider surgery. It would require that she stay in the hospital three or four weeks after the surgery and, as was the common practice, someone should stand by in case a sudden blood transfusion was needed.

In the spring of 1942, Sylvia was anesthetized and wheeled into an operating room. Her record doesn't specify which hospital, but it was most likely the Neurological Institute. A left frontal craniotomy was performed on Sylvia. A craniotomy was not a new surgery; in fact, it predates medical literature, but in modern times, it was used in cases of severe brain traumas or tumors. A hole was drilled in her head and an instrument inserted to cut out or sever the part of the basal ganglia that was triggering the tremors and uncontrollable movements. It was a brand-new technique for the treatment of Parkinson's patients, and a large number of those patients were suffering from postencephalitic Parkinsonism.

Sylvia did improve. Within a few months, she gradually began to walk again and feed herself with some assistance. The neurosurgeon recommended that family members and Sylvia's regular physicians do everything they could to encourage her independence. The surgeon ended his postoperative summary by writing that Sylvia was also presenting quite a problem for the nursing staff because of her "tendency to be incessantly demanding."

By then, the Second World War had broken out. Constable's husband, an army surgeon, was stationed in Memphis, Tennessee, so she went with him and joined a promising neurology clinic headed by Dr. Eustace Semmes, a student of famed neurosurgeon Harvey Cushing. But Constable continued to correspond with Sylvia and her family.

In spite of Constable's high hopes and the surgery's temporary success, by 1945 Sylvia's condition had worsened rapidly. She seemed to suffer a stroke, although Constable said it was most likely just the effects of the disease. Sylvia showed a "very marked tendency to childishness and irrational talking and thinking." She was completely bedridden, her feet badly deformed, her face distorted. Sylvia could barely talk above a whisper and cried most of the time. Within a few weeks, Constable received a handwritten note from Sylvia's sister.

March 20, 1945

My dear Dr. Constable—

This is to tell you that at last poor Sylvia is at rest. She died this morning, after having been in a coma for over a week.

We shall never forget your thoughtfulness and the many kindnesses you have shown Sylvia—you were surely one of her truest friends.

With deepest appreciation,
Sincerely Yours,
Margaret

CHAPTER 22

I Have Seen the Future

There was a sad sort of symbiosis between epidemic encephalitis and the city of New York. When the epidemic first broke out in the 1920s, the city had rallied around public health to fight it, while the success of Tilney's vision for New York neurology culminated in the greatest neurological center in the world. During those years, from the end of the war to the beginning of the Depression, New York itself had been in the midst of a great economic, technological, and medical boom.

The year 1930 marked the end of two brilliant, parallel tracks: one for neurological research, the other for the city of New York. That year, William J. Matheson died, stunting the work of the Matheson Commission and the Neurological Institute's hope for worldwide recognition and acclaim. If the Progressive Era had helped pave the way for the rapid and impressive medical research surrounding epidemic encephalitis, medicine's failure, too, was born of that era. That which did not succeed, progress, or grow quickly fell to the wayside. There was no solution to epidemic encephalitis, no answer. Neither Tilney, Jelliffe, nor Neal wrote about their disappointment personally, but it was undoubtedly present. A decade of work and hope, as well as close connections to patients, ended without resolution on any level.

And that same year, the Depression began, wearing away all the luster of the 1920s to expose a rusting and decayed city over the next ten years. The Depression changed New York dramatically. It imposed limits on a city that thrived on breaking boundaries.

Some changes in the city showcased the progress of the times, while others portrayed the decline. Automobiles still covered the streets, and by the 1930s, horses and buggies were almost entirely gone from the city. Gas stations became so necessary that the companies began issuing a card that could be used as credit to purchase fuel, although it would be years before true credit cards became commonplace in all stores.

Dining out meant going to soup kitchens and penny restaurants, though restaurants in general would make a strong comeback when Prohibition ended in 1933. Even a few of the speakeasies, like Jack and Charlie's '21' Club and the Palm, survived in New York. Now that people ate at home more, there was a return to homemade foods based on the seasons: jellied fruits, roast beef, five-minute cabbage, parsnips, sand tarts, and gingerbread in the colder months and cold meats, pork chops, wilted dandelion greens, beans, and fried tomatoes during the warm ones.

The Depression led to a major turnover in New York City politics as well. Mayor Jimmy Walker had typified the 1920s—he was also a product of Tammany Hall. Walker was colorful, a drinker and a gambler, and he became known as "the Mayor of Night." He was also negligent. He spent 143 days out of his first two years on vacation. The 1930s was not the decade for a mayor like Walker, and after the Depression hit, it was obvious how much city services had deteriorated. Worse, money funneled into New York for relief efforts quickly went into the pockets of Tammany politicians. Under pressure from New York's governor, Franklin D. Roosevelt, Walker finally resigned amid political scandal and moved to Europe in fear of being prosecuted.

In the next major election, in 1933, Republican mayor Fiorello Henry La Guardia came into office and vowed to turn the city around. As one historian said, "When La Guardia took office, city finances were chaotic, crime was rampant, housing was a mess, and soon the city's unemployed equaled the entire population of Buffalo." The situation was so desperate that state and federal governments had to supply 75 percent of the city's relief burden—it didn't hurt that La Guardia had a strong ally in Roosevelt and his New Deal administration. The U.S. government spent a billion dollars on New York City during the years 1933 to 1939.

Ironically, one municipal agency would suffer from progress: public health. As the Depression deepened and ordinary people were no longer able to pay physicians, the government created health insurance for the unemployed. While New York City's exemplary health department had at one time battled Tammany Hall and city politicians, it now faced federal politics. Municipal public health was no match for the New Deal. Under La Guardia and subsequent mayors, public health would receive less attention and fewer funds.

Not everything in New York was failing, however. John D. Rockefeller, Jr., started work on the Rockefeller Center, employing seventy-five thousand men in the ten years between 1930 and 1940. It was the largest private building project in modern times. The Music Hall built as part of the project was considered the most glamorous theater in the world, and in keeping with the popularity of radio, it was renamed Radio City Music Hall.

The Empire State Building, the tallest in the world, was erected during the Depression. As E. B. White wrote about New York, "It even managed to reach the highest point in the sky at the lowest moment of the depression."

In addition to focusing on the critical civil services needed during the Depression, La Guardia also understood the need to restore New York's pride, and one of the places he accomplished that was in Central Park. The park had been a political sore point for decades. There was a nonstop struggle between those who wanted to preserve its pastoral quality and those who wanted to make the park a place of urban recreation, with playgrounds, sporting fields, ice skating, zoos, and restaurants. La Guardia appointed Robert Moses as parks commissioner, and using a large chunk of the federal relief funds, they modernized the park, as well as the city.

During the early 1930s, Central Park acted almost as a reflection of New York City itself. Wealthy socialites visited the Casino, a private club frequented by Mayor Walker. With a black-glass ballroom and a terrace, it was considered the most elegant restaurant in the world, and coincidentally, its chef had been personal chef to the Rothschild family in London. In addition to the Casino, new hotels edged the park—the Essex House, the Sherry-Netherland, and the Pierre. Already well established on Central Park South was the Savoy-Plaza. Yet, only yards away, in the middle of the park, the Lower Reservoir had been filled and now served as a home to a number of squatters who built shacks out of scrap materials and called their community Hooverville. Much to the irritation of the residents along Central Park West and Fifth Avenue, Robert Moses also built more playgrounds and athletic fields, which brought in all classes of people and many immigrants who had nowhere else to spend their leisure time. By then, 70 percent of New Yorkers lived in upper Manhattan within an easy walk of the park.

Moses also had the Casino closed down, and as an alternative, a tavern was built on the Green and opened as a more affordable option—the sheep that formerly occupied the Green were sent to Brooklyn for convenience, as well as to protect them from hungry men growing desperate during those times. Moses had the menagerie built into an actual zoo. And, in an attempt to provide jobs to the jobless, hundreds of men were hired to clean the park, so it actually looked better than it had in years.

La Guardia set his sights on one more way New York could return to its stature as "the greatest city in the world." In 1935, in the depths of the Depression, the country was looking for a place to host the 1939–40 World's Fair. La Guardia and a number of New York businessmen decided New York was just the place. The plan showed a remarkable faith in the city of New York. After all, the city was home to the Stock Exchange, the starting point for the Depression and a glaring reminder of its tenure. A number of Wall Street businesses were under investigation for corruption at that point. The city itself had been so crippled by the economy that the federal government had to help restore it. Everything from the park system to housing to unemployment was in a shambles. Yet, in just four years, the city planned to host the World's Fair to prove not only to New Yorkers, but to all Americans, that the country would surface from the Depression. For New Yorkers especially, it gave them something they had not had since the autumn of 1930: hope.

The fair, aptly named "Building the World of Tomorrow," focused on the future, science and technology in particular. An enormous orb and a seven-hundred-foot spire stood as the great monuments to the fair, and the architecture of the buildings was Art Deco with modern, geometrical features. During the day, they stood out as fascinating models of architecture; by night, they glowed with a sense of things to come. Some of the modern inventions on display included a pencil sharpener, IBM's electric calculator, robotics, FM radio, fluorescent lighting, long-distance phone calls, even Wonder

Bread. Moving walkways took people through some exhibits where visitors could watch an actual car being assembled. And as tangible proof of this modern era, the opening of the fair was conducted in high definition on early model televisions by the National Broadcasting Company.

For two seasons in 1939 and 1940, 45 million people visited the fair. Countries from all over the globe participated—the only world power not to be part of the fair was Germany. Fatefully, when the fair closed, the four thousand tons of steel used to build the orb and spire were melted down to make weaponry for World War II. It seems science and technology were indeed the way of the immediate future.

After the fair, people kept a number of collector's items, including a button that read: "I have seen the future."

In 1942, Sylvia and all other patients involved in the Matheson Commission's work received a letter from Neal explaining that in July the "Matheson Commission's service to patients has come to an end due to the lack of funds to carry on this work." Neal went on to say that she wanted each of them "to be under the care of a neurologist who thoroughly understands the problems of patients afflicted with encephalitis. I am suggesting, should you need the services of a physician in the future, that you should consult Dr. Kate Constable."

During the 1930s, the Matheson Commission had included work on all types of encephalitis, including the St. Louis outbreak, among other neurological diseases. Preliminary and important work on multiple sclerosis was started by the commission as well. But the nation was entering war again, just as it was emerging from a deep economic depression. In addition to that, the Matheson Commission could no longer find acute cases. With the epidemic well over, its clinics were instead filled with the growing number of chronic patients. Many of the disease's researchers had moved on to new projects.

In a desperate plea to donors, Dr. Hubert S. Howe wrote that "ten years of painstaking effort toward a goal that might well be reached in the not too distant future will have been made almost in vain if the work has to be discontinued at this time." He believed that just the sum of $5,000 per year for the next five years would make the difference between success and failure "of a truly worthwhile endeavor." In his letter, he also mentioned the patients themselves, dependent not only on the clinic for medical help, but for hope: "it would be a real tragedy if they were one day told they could no longer return."

Tilney accepted a position as medical director of the Neurological Institute in 1935, but by then his health was poor. When he died of heart disease three years later, the physician who replaced him did not believe encephalitis lethargica to be an infectious disease and suspected that the entire epidemic had been misinterpreted by the medical community. In the coming years, it would be classified as a syndrome more than an epidemic. As medical historian Kenton Kroker has noted, "For American researchers, the temporal horizon of the disease no longer pointed toward the future of neurological research. Instead, it gestured toward the past as a novelty of history unlikely to repeat itself."

On that point at least, the physicians of the 1930s were wrong.

Dr. Josephine Neal continued to work with the health department, as well as the Neurological Institute, finally asking for cab fare to stop having to make the long trip from Gramercy Park up to Washington Heights by subway. She was aging, and the commute had become tiring. Nonetheless, Neal would continue her work until her death in 1955. By the time she died, at the age of seventy-four, this woman physician who had struggled even to be allowed into medical school had been the associate director, as well as head of the meningitis division, of the Bureau of Laboratories in

New York's Department of Public Health; a professor of neurology at the Columbia University; the director of the Matheson Commission; and a recipient of an honorary Doctor of Science degree from Bates College, a similar honor from Russell Sage College, and an Elizabeth Blackwell Medal for women in medicine. She had published definitive articles or books on meningitis, epidemic encephalitis, polio, and St. Louis encephalitis.

In 1938, Dr. Frederick Tilney died, at the age of sixty-three, after several months of heart disease. In Manhattan, a group of doctors, scientists, lawyers, businessmen, and socialites began raising money for what they hoped would be a $150,000 research fellowship in neurology in honor of Tilney. He had served as chairman of the Neurological Institute for fifteen years before becoming its director. Toward the end of his career he had focused on child delinquents and the criminal brain—in part due to the sudden rise in juvenile delinquency after the sleeping sickness epidemic. And, of course, he had been a researcher and member of the Matheson Commission. Tilney was considered by his peers to be one of the most distinguished and active figures in American neurology.

Slowly, the great pioneers of American neurology began to disappear, and neurology itself was being overshadowed by the exciting new field of neurological surgery. Neurosurgery, and in particular psychosurgery, was the new medical frontier. As a result, neurology and psychiatry, two fields inextricably linked to epidemic encephalitis, would split once and for all. Even the term "neuropsychiatry" would quietly drop the prefix "*neuro*," and as one medical writer observed, "In a startling reversal of roles, the senior and previously more powerful field of neurology now played second fiddle to its upstart cousin."

One of the greatest proponents of neuropsychiatry, Smith Ely Jelliffe, who had once called it "the fairy godmother of medicine," died on September 25, 1945, at the age of seventy-eight, living just long enough to see the end of World War II the month before. His savings had been wiped out in the stock market crash of 1929, so he had continued working right up to his death. He sold the bulk of his book collection—fifteen thousand books—to the Institute for Living in Hartford. In spite of spending years on the outside of neurology, Jelliffe was celebrated in 1938 at a symposium at the New York Academy of Medicine, attended by Tilney just months before his death. Five hundred others joined Tilney in honoring Jelliffe and the thirty-fifth anniversary of the *Journal of Nervous and Mental Disease.*

As a full-fledged psychiatrist, Jelliffe was also a strong objector to psychosurgery. In the late 1930s, he attended a medical meeting in which a study on the new procedure known as a frontal lobotomy was presented by Walter Freeman and his partner James Watts. Jelliffe compared psychosurgery to burning down the house in order to roast a pig. In particular, it bothered psychoanalysts that surgeons permanently destroyed the one part of the brain that therapy might heal. Other psychiatrists likened it to "partial euthanasia." Throughout the 1930s and '40s, they would wage a losing battle against psychosurgery and lobotomies until the advent of antipsychotic drugs in the 1960s. One of the early lobotomies performed by Freeman and Watts was that of Rose Marie Kennedy, whose low IQ and behavioral disorders proved unmanageable to her father. The operation did not go well, and she was left completely incapacitated, incontinent, and unable to walk or talk for the rest of her life. Another famous lobotomy case was that of Rose Williams, playwright Tennessee Williams's beloved sister. While Williams was away at college, his parents had his sister, diagnosed with schizophrenia, lobotomized. As in the Kennedy case, Rose Williams was left in much worse condition. The tormented Tennessee Williams never forgave his parents, and he based several of his characters, such as Blanche in *A Streetcar Named Desire*, on his sister Rose.

In spite of early reports of success with lobotomies, a darker side to the mutilating surgery began to emerge. Patients were considered improved because they were so placid and lethargic—nurses had trouble even moving them. They also returned to a childlike state in many instances. The "success" of the lobotomy then was that it silenced raving, difficult patients. With initial excitement of psychosurgery fading fast, the majority of neurosurgeons had already turned their attention to herniated discs, tumor surgeries, and brain injury.

What's more, by the 1960s, drugs and shock therapy could accomplish the same thing psychosurgery could. Neuroleptics like chlorpromazine (Thorazine) produced what were called "chemical lobotomies." They were originally marketed as miracle drugs, and the side effects were downplayed. The "therapeutic" effect of these drugs was that they damaged the basal ganglia, the same part of the brain injured by encephalitis lethargica. Patients became essentially tranquilized and lethargic, and wore masklike, waxy expressions. In addition to this, because the drugs hurt the areas of the brain that control movement, they began suffering from Parkinson's. In an ironic turn of fate, psychosurgery and neuroleptics, in their quest to cure mental illness, returned patients to a condition almost identical to the one neuropsychiatrists of the 1920s had worked so hard to conquer.

There is still no answer as to why the encephalitis lethargica epidemic was so widely forgotten. It may have been that the 1920s was a decade in which people wanted to enjoy life, not be reminded of its fragility. Or perhaps the disease, like World War I and the 1918 flu pandemic, remained an open wound in American memory. With so many advances in medical science, epidemic encephalitis may have been overshadowed by the medical successes. Or maybe the despair of the 1930s simply eclipsed it. One reason the disease was forgotten is certain: the thousands of survivors that epidemic encephalitis maimed ended up in institutions; they were removed from everyday life and, consequently, the collective memory of society.

In his book *America's Forgotten Pandemic: The Influenza of 1918*, Alfred Crosby addressed reasons why the 1918 flu pandemic was almost lost to history: "The very nature of one disease and its epidemiological characteristics encouraged forgetfulness in the societies it affected. . . . If the flu were a lingering disease, like cancer or syphilis, or one that leaves permanent and obvious damage, like smallpox or polio, America would have been left with thousands of ailing, disfigured, and crippled citizens to remind her for decades of the pandemic." The sleeping sickness epidemic did leave thousands of ailing, disfigured, and crippled citizens. And still it was forgotten.

The victims of encephalitis lethargica may have been overlooked to a large extent, but they still existed in the decades following the epidemic. In 1946, for example, Mayor William O'Dwyer, New York's one hundredth mayor, lost his wife after a fifteen-year struggle with chronic encephalitis symptoms. She had the typical onset of Parkinsonism, and by the time she died, she was confined to a wheelchair in Gracie Mansion. She passed away in the mayor's first year in office.

Two decades later, Dr. Oliver Sacks found several of these mysterious victims of the epidemic at the Beth Abraham Hospital—although he added that by the 1960s there was no major country in the world *without* postencephalitic patients. Sacks was a young neurologist when he encountered these relics from a past epidemic. Piecing together their medical histories, he was able to find the one similarity among these frozen patients: they were all survivors of the sleeping sickness epidemic. His experiments with the L-dopa drug in 1969 proved to physicians and family members that the survivors were in no way extinct. Up to that point, most doctors believed that the "virus" of epidemic encephalitis had damaged the parts of the brain that think as well as those that control movement. Believing that the survivors of the epidemic, these breathing statues, seemed unaware must have offered some small consolation. Once the patients awakened, however, many were able to describe exactly what it felt like to be trapped in their bodies by physical dysfunction, but still very much alive inside.

Epidemic encephalitis was considered one of the most important diseases in the development of twentieth-century American neurology. In all, it afflicted an estimated five million people worldwide, killing one-third of them and leaving one-third to die inch by inch, minute by minute in asylums. One neurologist wrote that no other infectious disease affected so large a portion of its victims, or for so long a period of time.

In June 1942, the *Journal of the American Medical Association* reported that, "The common worldwide and devastating disease encephalitis has become familiar to every physician." And in 1986, one medical journal called encephalitis lethargica "a disease of momentous importance for three decades." The very neurologist for whom the disease was named, Constantin von Economo, had predicted it would never again disappear from medical memory: "One thing is certain: whoever has observed without bias the many forms of encephalitis lethargica . . . must of necessity have quite considerably altered his outlook on neurological and psychological phenomena. . . . Encephalitis lethargica can scarcely again be forgotten." By 1935, however, nearly all medical literature on the subject stopped, and by 1960, encephalitis lethargica was no longer even taught in medical school.

When Oliver Sacks encountered his *Awakenings* patients, these lepers of the twentieth century, as he described them, he wrote, "I would not have imagined it *possible* for such patients to exist; or, if they existed, to remain undescribed."

The tragedy of epidemic encephalitis then was not the disillusionment of American neurology, nor the decade-long siege on

New York City and its public health system, nor even the failed efforts of the medical investigators; it was the slow extinction of its survivors. The city of New York survived the tumultuous times, and the epidemic became little more than a fleeting memory. Cities rarely immortalize their failures, and this was not just New York's failure, it was Philadelphia's, Chicago's, Boston's, Vienna's, London's, Paris's. It was also medicine's failure. When the money to the Matheson Commission stopped, and neurology shrank behind the larger interests of psychiatry and neurosurgery, hope vanished. The neuropsychiatrists who had spent over a decade trying to understand this disease died one by one. The chance was gone, and medicine moved on.

The survivors of epidemic encephalitis were condemned to a life in mental asylums—institutions that were deteriorating so quickly they would no longer exist by the 1970s. What's worse, those patients knew they had been forgotten. Their brains and bodies may have been damaged, but their minds remained lucid. It's hard to imagine which would be worse—never knowing how or why they got this debilitating disease, or knowing they had been left in an asylum and forgotten as a result of it.

It wasn't until 2002, when one of those patients was discovered after seventy years in an institution, that encephalitis lethargica came to the attention of modern medicine once again.

CASE HISTORY SEVEN

London, 1931 and 2002

NAME: Philip

PHYSICIANS: Multiple

CHAPTER 23

Philip

Magton was a land where children were always kept safe. It was a Neverland of sorts. Philip used to imagine that he flew out his bedroom window over Magton Heath, the ground purple with fresh-blooming heather, the wan evening light settling over a land where fairies perched on tree limbs. "Remember that nothing can hurt you in fairyland," Philip scrawled in a palm-sized leather notebook.

Imagination had always been a gift of Philip's, but he had others as well. He had been playing the piano since the age of three and composing his own tunes. Philip's parents, especially his mother, were sure he would grow up to be a composer or pianist. He was even considered something of a child prodigy. While other children struggled to piece together puzzles, Philip bored of them, turned them upside down, and put the pieces together without the images. He also built entire cities in his attic, made from bricks and wooden cotton reels.

It was an idyllic childhood in Crosby, England, a residential neighborhood outside of Liverpool, where Philip's father was a silversmith. On sunny days, Philip and his sister played in the garden and stole loganberries to eat. And on winter days, Philip wrote his stories, the otherworldly tales of Magton, tinged with magic, draped in the tapestry of childhood imagination.

Then, something inside Philip's mind shattered.

The year was 1931, and for Philip time stopped; the world seemed to cease moving, imploding. Philip's parents and his sister appeared by his bedside, and in between fits of sleep, he could comprehend some of what they said: fever. Flu. Not recovering. Lethargy. Sleepy. A disturbing lilt in their voices could be perceived—it was the way they asked. The fear they gave off. The feeling that this mysterious disease was out of their control.

In nightmares, Philip was trapped inside his own body. It was worse than death; it was purgatory. Though his body would not respond, his mind was always awake, filled with nothing. An overwhelming, deafening, suffocating nothing.

For his parents, the transition was terrifying. Philip would not wake, even when shouted at, even when shaken. When he did open his eyes, they stared, frozen, at something no one else could see. The day that his illness finally broke, and Philip began to recover, the whole family felt enormous relief. But the Philip who awakened from that deep sleep was not the same.

At first, the differences were small. Philip's demeanor could become clouded with a darker sense of mischief and unruly behavior. Teachers complained. Then he began walking with an unusual gait, his shoulders hunched forward. His physical appearance began to change, and he would repeat the same movements over and over again, obsessively. Finally, one night, when playing a favorite game of hiding beneath his bedsheets, he refused to come out. He stayed there for three days.

Philip's parents took him to see a leading neurologist—Dr.

George Auden, poet W. H. Auden's father. He told Philip's parents the sober news: Philip suffered from the effects of encephalitis lethargica. He was admitted to the only place available at the time, a mental institution in a ward with senile, elderly men. Philip would live there for the next seven decades, enduring mind-altering medications and shock treatments. By the year 2002, much too far in the future for any of the doctors or their patients to imagine at that point in time, Philip would be the world's last known survivor of the encephalitis lethargica epidemic. Surely the doctors during the epidemic believed that a cure or a vaccine would arrive in the following decades. With so many medical advancements ahead of them, including electron microscopes, CT scans, MRIs, and mapping of the human genome, surely the answers to this epidemic would be found.

That would not be the case. And so when Philip died later in 2002, his sister gave to medicine the only thing science could never fix—his brain.

CHAPTER 24

Gray Matter

Sliced and cross-sectioned, a human brain looks more like a sea creature from the lightless depths of the ocean than the complex organ that it is. It is flat, putty-colored, and shaped like an inkblot test.

Dr. John Oxford, who describes himself as a virus hunter, was in London the day Philip died, so he took a train to the institute and was unceremoniously handed Philip's brain in a box. At Oxford's lab, the brain was sliced and fanned out on a metal sheet. Oxford has long believed there to be a connection between the 1918 flu pandemic and encephalitis lethargica. Oxford has more than a passing interest in the flu. As a young medical student, he studied under one of England's famed virologists, Sir Charles Stuart-Harris, a researcher who served on the team that isolated the first flu virus. Oxford continued his investigations of the flu in the decades to come, even conducting research on a vaccine for the 1968 flu outbreak. Oxford's theory that encephalitis lethargica is somehow related to the 1918 flu is a strong one.

Josephine Neal, in her writings about the disease, noted that a majority of children's cases of epidemic encephalitis were preceded by flu. In the United States, encephalitis lethargica cases were highest in the cities where pandemic flu had hit hardest, primarily Philadelphia and New York.

Another compelling connection between the 1918 flu and encephalitis lethargica is the story of American Samoa. Roughly one hundred miles away, neighboring Western Samoa was devastated by the flu pandemic and encephalitis lethargica, losing 22 percent of its population. American Samoa therefore quarantined itself, avoiding the influenza pandemic entirely and suffering only two cases of encephalitis lethargica.

And, after all, there seems to be a logical connection when history's most devastating flu pandemic struck at the same time as history's largest pandemic of sleeping sickness.

Many doctors at the time of the epidemic also believed the two were linked—that the virus left a smoldering infection that later flamed up into encephalitis. "This is a major, unsolved mystery," Oxford told the BBC. "My personal interest in this project is in hopefully discovering, hopefully identifying, the cause of that mega-outbreak in the '20s, so we can help prevent such a thing happening in the future."

Today, cases of encephalitis lethargica still happen sporadically throughout the world. The disease has never disappeared entirely, but it has not again occurred in epidemic form. There has even been a fund established in England and named for Sophie Cameron, a promising young student who planned to be a doctor, but instead fell into a deep sleep from which she never recovered. The Sophie Cameron Trust raises money for further research on encephalitis lethargica. In Britain especially, physicians have been focusing on these outbreaks.

The BBC's 2004 documentary on encephalitis lethargica, "Medical Mysteries: The Forgotten Plague," shows what it's like to watch a normal, healthy person go quickly insane or quietly immobile. In one modern segment, a young woman's case was frighteningly like those of the 1920s when she flew into a sudden rage, threw furniture, and became truly unrecognizable to her parents. In her hospital bed, her body tensed, she panted, her hands clawed, and even her toes curled tightly. The hospital staff had never seen anything like it, and her physician worried that the brain swelling would soon cause brain damage—what patients in the 1920s would have experienced as chronic symptoms. The young woman eventually recovered after receiving large doses of steroids, and after her recovery, her doctor showed her video footage of herself during the illness.

"This doesn't feel like me," the young woman said. "It just doesn't feel like I'm watching myself . . . it's amazing."

Though steroids do not work in every encephalitis lethargica case, they do work in some, and that offers new clues to the disease. It is very likely that the disease is actually an autoimmune response. That means that some pathogen enters the body, and the human immune system makes a mistake, attacking the body along with the intruder. Most often, the body responds by growing inflamed—in this case, swelling in the brain. Encephalitis has often been culpable in immune responses to infections caused by viruses or even the vaccines intended to control them. Viruses like measles, syphilis, herpes, HIV, Epstein-Barr, and several arthropod-spread ones like West Nile cause swelling and neurological complications. An autoimmune response might also explain why the severity of the infection had little correlation to the encephalitis—a mild case of the flu could cause encephalitis the same way a severe case would because it was the immune response, not the pathogen, that was to blame. And autoimmunity would explain why some patients have responded well to steroids, which decrease swelling and slow the immune response.

It's possible then that had patients in the 1920s had the benefit of steroids during the acute stage of their disease, their brains might have been spared the damage that eventually caused so many chronic disabilities. Steroids may even have helped the patients who died while still in a deep sleep.

Like those during the epidemic, however, today's cases are erratic and difficult to predict. In 2006, a healthy high school football player was admitted to a hospital in Houston, Texas. He had experienced some sort of seizure at school, his body growing tense and his eyes rolling into the back of his head, before he finally lost consciousness. Eventually he regained consciousness, but he could only say that he had felt numbness and tingling on one side of his body earlier that week. He showed no other symptoms.

The hospital ran tests, including a CT scan, but the doctors failed to find anything out of the ordinary. Later that afternoon, the boy told his mother his food didn't taste right. He felt tired and lethargic, and his mother noticed that he had started making an unusual sound—a hissing noise. His body soon twisted left and grew stiff. His doctors struggled to find out what was causing these unusual symptoms and how to control them. Physicians suspected some type of encephalitis, but his MRI scans continued to come back normal.

Over the next several weeks, the high-schooler lost thirty pounds and grew more lethargic and childlike. At times he experienced psychotic episodes; at other times he had outbursts of obscenities. His mother barely recognized him, describing the situation as similar to some sort of demonic possession.

One of the physicians at the hospital had an interest in medical history and had happened upon an article about Constantin von Economo and encephalitis lethargica. The physician ran through the list of symptoms, most of which he had seen in this case. The boy could fall asleep for days at a time, and it was difficult to wake him, even for food. With no treatment options available, the hospital tried both steroids and levodopa. The steroids and L-dopa had no immediate effect on him as the physicians had hoped they would, and so they put him in a protective environment and waited to see what happened. Slowly, over the next few weeks, the patient did recover. To the boy's bewildered parents, the physician explained how rare encephalitis lethargica is and added, "We've just scratched the surface in understanding this form of encephalitis."

The question nettling Oxford and other neurologists today is the same one that confronted the neurologists in the 1920s: what is encephalitis lethargica's relationship to the flu? The question is not just a matter of history; it has a direct effect on our future. If most sleeping sickness epidemics have coincided with flu epidemics, as many scientists have found, the next major flu pandemic could again produce these tragic encephalitis patients.

There seems little doubt that an avian or swine flu will strike once again. Either an entirely new flu strain will overwhelm human populations or one of the strains already identified will mutate into a more deadly form. At that point, influenza will spread around the world like a flame chasing oil. In 1918, that was accomplished through troop movements during World War I; today, it would be spread by mass transportation. In 2009, we saw just how quickly the H1N1 flu spread through airplane travel, and in spite of quarantines, how far it stretched.

Oxford's modern hunt for the virus began with an amazing discovery from the past. In the basement of his hospital, Queen Mary's in London, he found shelves stacked with wooden boxes marked *Post-mortem*. The hospital had taken samples from fatal disease cases dating as far back as the turn of the century. He pulled the crate labeled *P.M.—1918*. In it, he found eight brain samples of patients who died of encephalitis lethargica during the epidemic.

Physicians during that time were unable to solve the mystery, and now in the twenty-first century, Oxford had found this brain tissue, preserved in formalin and sealed in wax as though the scientists of the 1920s were keeping a time capsule for scientists of the future. The pea-sized samples were sliced paper-thin and tested for genetic evidence of influenza using today's advanced technology. The tests came back negative. Likewise, tests showed that Philip's brain did not have any traces of the flu. And in the United States, a 2001 study conducted at the Armed Forces Institute of Pathology also studied archived brain tissue for the flu virus. Those results also came back negative.

This disappointing outcome left Oxford in the same position so many physicians found themselves in during the 1920s—with far more questions about this disease than answers.

There are other problems with the flu theory as well. Without a test to identify flu viral strains in the 1920s, how many cases were actually flu and not some other type of pathogen? It is also impossible to track with certainty how many flu cases actually sparked a case of epidemic encephalitis, or how many were mild enough to go unnoticed. In her book *Flu: The Story of the Great Influenza Pandemic of 1918 and the Search for the Virus That Caused It*, Gina Kolata pointed out the statistical conundrum: "Most people alive in 1918 got the flu. Even if encephalitis lethargica had nothing to do with the flu, by chance alone most people who got encephalitis lethargica would also have had the flu." And as American encephalitis lethargica researcher Dr. Joel Vilensky recently noted, "it is extremely difficult scientifically to prove a negative, in this case, to prove that influenza did not cause EL [encephalitis lethargica]." He wrote an article in response to a recent book published in Britain that unequivocally links influenza and epidemic encephalitis. Vilensky, however, has his doubts.

Vilensky, a professor of anatomy and cell biology at Indiana University School of Medicine, Fort Wayne, is working with other researchers on a database system, funded in part by the Sophie Cameron Trust, recording key symptoms of historical and contemporary cases of encephalitis lethargica. He hopes the database will not only give a more accurate estimate of the number of cases during the 1920s, but also help physicians identify modern-day cases as they arise. In the event that encephalitis lethargica *does* return with the next flu epidemic, Vilensky believes, "Difficulty in the diagnosis could again become a problem, and possibly a serious public health issue."

The difficulties found by Oxford and other physicians who subscribe to the flu theory of encephalitis lethargica have given momentum to the other popular current-day theory: strep throat. A young pediatric neurologist named Russell Dale was working at the Great Ormand Street Hospital in London in 2002 when he began seeing children with remarkable and unusual symptoms. Their parents, echoing the ones in the 1920s, said that it was as if the children had changed personality overnight. Dale consulted other neurologists, physicians who had been practicing for thirty or forty years, and none had seen patients like these.

"Encephalitis lethargica wasn't really taught in medical schools anymore because it was thought it disappeared," explained Dale, but having read Oliver Sacks's *Awakenings*, he developed a curiosity about the long-forgotten disease.

Today, Dale is a senior lecturer at the University of Sydney, Australia, and works with a children's clinic, specializing in brain inflammation. Over the last several years, he has seen twenty-five pediatric cases of what he believes is encephalitis lethargica. Though it's difficult to know if it is the same disease that circulated in the 1920s, Dale's diagnosis is based on three classic symptoms: movement disorders, psychiatric symptoms, and sleep disorders. No other neurological condition shares that exact symptomology.

Rifling through current and historical files, Dale has found a common link among the patients: sore throats. And like the flu, sore throats are more common during the winter months, when encephalitis lethargica usually peaked. Most sore throats are caused by simple bacteria called streptococcus group A, which appear like a chain of round beads in the blood. More than sixty types of streptococcus cause a wide range of diseases, like strep throat, tonsillitis, pneumonia, infections in teeth, and even a flesh-eating disease on the surface of the skin. Harmless chains of streptococcus also are found in humans, on skin, and on surfaces.

To substantiate his theory that streptococcus is somehow linked to encephalitis lethargica, Dale went back to von Economo's original studies in 1916 and 1917 and found that most cases had started with a sore throat. What's more, von Economo, in his own research, tried injecting bacteria into rabbits and was successfully able to produce an encephalitis lethargica–like disease. The type of bacteria he used was called diplostreptococcus—the type of strep that causes pneumonia.

Dale's theory expands upon the idea of autoimmunity, helping to explain *how* encephalitis lethargica happened, if not why. All bacteria have proteins along their surface that interact with the immune system. And the surface proteins of the streptococcus A bacteria look very much like the surface proteins of brain cells. In evolutionary terms, Dale explained, *all* cells are similar in their makeup. Human and ape DNA is almost 99 percent alike; we even share a close genetic relationship to rats, which makes them a popular choice in lab experiments. Like every other type of flora or fauna, bacteria also are living organisms, so our cells can closely resemble those as well. When the body comes under attack from any foreign agent like a virus or bacterium, it builds antibodies to fight the invader. But if the surface of the bacterium is similar to the surface of brain cells, the antibodies might not be able to tell the difference—a molecular case of mistaken identity. This "cross-reaction" confuses the body, which not only begins producing too many antibodies, but attacking the brain cells as well as the pathogens. And, according to Dale, the basal ganglia are particularly vulnerable to antibodies attacking the brain.

Eventually, when so many pathways going to or coming from the basal ganglia are damaged, communication to the parts of the brain that control both movement and emotion become erratic. The misfired messages might tell the body to move frantically, quickly, causing shaking and repetitive motions. Or the flood of messages might do just the opposite and neglect to tell the muscles to move at all. In much the same way, mixed messages sent to the frontal lobe might very well affect the personality of the patient.

In one study, Dale tested the blood of some twenty new cases of encephalitis lethargica. In most, there had been a preceding throat infection, and in nearly all of them, Dale found antibodies that react against the basal ganglia. Dale hopes that treatments will be able to target the problem by suppressing the immune system and essentially telling it to slow down the overproduction of antibodies.

"The idea of infection leading to a brain disorder is very old," said Dale in one interview. After all, medical literature during the epidemic itself was peppered with references to not only the flu triggering a brain infection, but also strep bacteria. In addition to von Economo's experiments with the strep that causes pneumonia, Rosenow, in the 1930s, was following the type of streptococcus responsible for infections of the teeth. And *Science* magazine reported in 1933 that a sample of streptococci had been taken from an encephalitis lethargica patient at St. Elizabeth's Hospital for the Insane in Washington.

Like Oxford's theory, Dale's, too, leaves questions, the most obvious one of all: why did encephalitis lethargica occur in an epidemic form?

Regardless of whether or not that question is ever answered, even the few cases seen today are tragic enough in their own right to deserve the attention of modern medicine. As Dale, who has watched two dozen children succumb to encephalitis lethargica, says, "It is devastating."

CHAPTER 25

Past or Prologue?

The answer to the great sleeping sickness epidemic may very well lie in a combination of today's theories. If encephalitis lethargica is, in fact, an immune response, the human immune system provides the most clues. One consistent characteristic of the encephalitis lethargica epidemic was that it struck concentrated populations, often those where other diseases were already circulating, like army camps, field hospitals, and crowded metropolitan areas. Where people are concentrated, so are germs. Perhaps, then, it wasn't one single pathogen that was to blame, but many.

For encephalitis lethargica to occur as an epidemic, it had to be shadowing one or more infectious diseases during the 1920s. The flu pandemic was certainly the largest of the time, but even if there is a connection to influenza, it may only be a *relationship*, not a cause and effect.

As it turns out, the immune system is also to blame for the high mortality in the influenza pandemic in one of two ways—either by working too hard or not working hard enough. The 1918 flu pandemic was set apart from other, less deadly outbreaks of flu by the shift in pattern. Instead of killing the very young or the very old, it homed in on the healthiest portion of the population. And, often, it was those healthiest men and women who died within hours. The reason for this may have been their strong immune systems. If the standard response to an invading virus is fever and fluid secretions in the lungs, healthier victims with a good immune system might overreact—what scientists today broadly refer to as "cytokine storm." If so, the healthiest and strongest immune systems would produce so much fluid, so fast, that the patient could literally drown.

It is more likely, however, that the 1918 flu pandemic was so deadly because of weaker immune systems, not stronger ones—and there we may find some answers to the encephalitis lethargica epidemic. Most people with a case of pandemic flu—without complications—survived. Those who died fell victim to a secondary infection, the most common being pneumonia. That would also explain the unusual W-shaped mortality pattern of the flu victims—high deaths among infants, high deaths among twenty-five- to forty-year-olds, and high deaths among the very old. It was the spike among young adults that was out of the norm, and the reason was probably their lifestyle. After a case of the flu, their immune systems were weakened, but most young adults, like soldiers, parents, or workers, went right back to work and were frequently and consistently exposed to contagious germs, which took hold like burrs in their weakened immune systems. What's more, the flu virus had already destroyed cells throughout the respiratory tract, making it an ideal place for bacteria to flourish.

In 2008, a study aimed at protecting public health during the next flu pandemic showed that pneumonia was the leading cause of death during the Great Flu and went so far as to use archived brain samples from 1918. All had evidence of damage from bacteria. And one of the leading bacteria to cause pneumonia: streptococcus A.

So far, only the strep that causes a sore throat has been linked to encephalitis lethargica, but different forms of strep appeared prolifically during the epidemic. In World War I, strep, as well as staph bacteria, were rampant in battlefield wounds, as well as in the lungs of flu patients. And the 1920s saw an epidemic of scarlet fever, caused by yet another type of strep A, which killed thousands of people in New York. In 1921 alone, there were over thirteen thousand deaths from scarlet fever, more than double the number of deaths from the year before. Bacterial meningitis was also occurring in epidemic form.

What's more, strep was only one of the many deadly pathogens of the 1920s. Health reports are peppered with cases of polio, measles, diphtheria, septic sore throat (not the same as strep), typhoid, and tuberculosis. Troop movements helped spread epidemics faster and more efficiently. This was also the height of the vaccine age. For the first time, vaccines and antitoxins were being used widely in the population—not only for viral diseases, but for infections like diphtheria and meningitis that would be treatable with antibiotics decades later. These vaccines may have exhausted the immune system. The environment in which people lived also changed, with cleaner water and food in most major cities. Sanitation had improved. As polio proved, the cleaner the world became, the less immune and more vulnerable people became.

In the complex world of epidemiology, it could have been the overwhelming number of diseases—what one physician called satellite infections—circulating at the time, from a highly virulent flu to strep throat, from scarlet fever to pneumonia. Perhaps people's immune systems were just overtaxed, making them more sensitive to infection and making a defeated immune system more likely to confuse the flood of foreign cells with the body's own cells. Somewhere on the evolutionary ladder of disease resistance, we may have stepped back a rung or two, and as a result, encephalitis lethargica got a foothold.

When and if this mystery is solved, it will certainly be one of medicine's longest and greatest examples of epidemiological prowess. The medical investigators who studied the disease during the epidemic are long since gone. Nearly a century later, a new set of medical investigators is building upon those old case studies, circling the same theories, trying to find a pattern in a disease with no clear archetype.

With the benefit of hindsight, we now know the epidemic of encephalitis affected as many as five million people worldwide, but scientists also speculate the number may be lower. We also know the epidemic probably started earlier than 1916, with sporadic cases as early as 1903 and an outbreak in Bucharest in 1915. The major epidemic came in two waves—the first began in 1916 and peaked in 1920. After that, the epidemic waned to just a few thousand cases worldwide. The second wave hit suddenly in 1924, and the number of cases rose dramatically, especially in the United States. And then chronic symptoms left a cruel legacy for several decades.

The hope is that many of the remaining questions will be answered before encephalitis lethargica returns in any type of widespread form. If, indeed, the flu and bacterial infections were tandem culprits in the pandemic of an autoimmune disease called encephalitis lethargica, answers are needed before the next major flu pandemic. Antivirals, antibiotics, and steroids could all help a patient recover *before* encephalitis has a chance to cause damage. Scientists are committed to that work today.

But the interplay between living organisms—both the ones causing a disease and the ones trying to survive it—has always been tenuous. Pathogens change and grow stronger, and the antibiotics and antivirals used against them lose their advantage. That fact will always make epidemics a frightening and real possibility.

The original epidemic, and the lesson learned from it, was too easily dismissed, too easily forgotten, and in the words of Oliver Sacks, "such forgettings are as dangerous as they are mysterious."

EPILOGUE

Virginia and the Forgotten Epidemic

There is nothing to save, now all is lost,
but a tiny core of stillness in the heart
like the eye of a violet.

—W. H. AUDEN, SON OF
DR. GEORGE AUDEN

With the epidemic of encephalitis lethargica so widely forgotten in the United States, it is a wonder that the subject ever came to my attention. Sleeping sickness may have been "the forgotten epidemic" to most, but for me it was always present in my grandmother Virginia. The prologue touched on her story. The year was 1929, and she was sixteen years old, living in Dallas, Texas. She had an infection of some type and was feverish the day she fell down the staircase of her parents' home.

Her parents put her to bed, and she fell asleep—the long, frightening sleep so many patients in this book experienced. She did not open her eyes for 180 days, missing most of her school year. In the days before modern diagnostic tools, her pulse grew so shallow that she was declared dead three different times by her doctors.

And then, miraculously, she awoke. A third of all sleeping sickness patients *did* recover. Physically, she remained weak and was never able to return to school. Thankful to have her awake again and recovering well, her family did not immediately notice the changes in her personality.

Virginia went on to marry and have four children, but she was never normal. She seemed distracted and lost, and it was whispered among family members that she had been that way since the great sleeping sickness epidemic. She was described as mentally "touched." It must have been frustrating for her family in an era that did not acknowledge or treat mental illness openly. Indeed, her parents may never have even known the extent of her affliction. The neuropsychiatrists of the 1920s would have found an injured brain and a broken mind, but the only doctor Virginia saw, a family practitioner, was in no position to diagnose chronic symptoms of epidemic encephalitis. It would not have mattered if they had known—nothing could be done for Virginia or any patients like her.

Had Virginia's mind been more intact, she most likely would have been diagnosed with severe depression. As it was, she did not even seem aware enough of her own thoughts or feelings to be cognizant of her unhappiness. Her mind was but a shade of the color it once was. Oliver Sacks described it best in *Awakenings*: "One can only say that most of the survivors went down and down, through circle after circle of deepening illness, hopelessness, and unimaginable solitude, their solitude, perhaps, the least bearable of all."

By the time I knew Virginia, it was five decades after her long sleep. She was thin, with a shock of black hair that silvered over time. Her eyes were as dark and glossy as blackbirds, and they always seemed fixed on something in the distance. She was a woman I had known most of my life, but felt I barely knew.

Virginia's behavior at times was almost humorous, the same way senility in its early stages can seem odd but harmless. At other times, it was just sad. She read newspapers and worked crossword puzzles, but left her apartment only to go to church or the beauty parlor.

She drifted in and out of conversations without warning. She would look at a dark wall for hours, the ash on her cigarette sifting off in soft gray pieces.

It could have been much worse. She could have been among the third of patients who died, or even worse, the third who spent their lives in institutions, damaged beyond repair. I had always assumed Virginia was lucky to survive her case of sleeping sickness and live a normal life. It was not until writing this book that I realized she probably did spend her life somewhere else, in that distant place only she could see, if not imprisoned physically, then mentally. What's worse, she always felt alone, but was not. Others like her survived and many in much worse condition. History let them go, and medicine remained asleep for far too long on the subject.

Virginia died in 1998. I was not living near her, so we spoke on the phone the night she went into the hospital, just hours before she passed away. I tried to memorize details about my grandmother—the smell of her lotion, the way her wrinkled hands felt like tissue paper. Knowing it was our last conversation, I also tried to find some connection, something to keep and remember—but there was only the hollow place sleeping sickness had left behind.

ACKNOWLEDGMENTS

I have two people to thank for inspiring me to write this particular book. My grandmother Virginia Thompson Brownlee survived sleeping sickness—and though she did not live to see the publication of this book, my hope is that she would find it an honest and unflinching portrayal of the disease. And, second, I want to thank Dr. Oliver Sacks. Without his work at Beth Abraham, the story of this epidemic would be buried even deeper in the archives of medical history, and the most poignant element—the patients—would never have been able to bear witness to the tragedy. I also want to thank Dr. Sacks for taking the time on a frigid morning in New York to meet with me and talk about this long-forgotten epidemic, and his assistant, Kate Edgar, for all of her help.

There were many historical collections I relied upon when researching this book. In particular, I would like to thank Stephen Novak and Henry Blanco in the Archives and Special Collections at Columbia's Augustus C. Long Health Sciences Library and Arlene Shaner in the Rare Book Room of the New York Academy of Medicine. Without their dedication to medical history, stories like this one could never be told. I would also like to thank Steve Weber at the Kings Park Museum for not only giving me a tour of Kings Park, but also sharing many details and photographs.

I want to offer my gratitude to the physicians and medical historians still dedicated to understanding this mysterious disease. In particular, I would like to thank Dr. Russell Dale for taking the time to answer questions and read certain passages from the book. I would also like to thank Dr. Joel Vilensky for many e-mail exchanges on the subject and for sharing documentary footage. And I want to thank Dr. Kenton Kroker for his thorough research and excellent article about the epidemic in New York and its effect on American neurology.

Very special thanks to David Everett, Associate Program Chair of the Master of Arts in Writing Program at the Johns Hopkins University. He read and edited an early draft of the manuscript. His thoughts and suggestions were invaluable to me, and the book is much improved for having crossed his desk. Ten years ago, David was my thesis adviser in the program and working with him again reminded me how much a writer can learn from a great teacher.

I also want to thank my agent, Ellen Geiger, and my editor at Berkley, Natalee Rosenstein, for their continued loyalty and confidence in me. And special thanks to Michelle Vega for her tireless effort and for answering endless e-mails from me.

Many friends and family members have offered their encouragement and friendship during the last few years. I want to thank personally Allison and Andy Cates, Margaret McLean, Tessa Hambleton, Lauren Kindler, Claire Davis, Jennifer Fox, Davida Kales, Elizabeth Levine, Candice Millard, Amanda and Cameron Jehl, and Temple Brown. I would also like to thank Holland Farrar, Tomas and Anne Ruiz, and Mark Crosby for entertaining me during my many research trips to New York. Special thanks to Mark for accompanying me on some of my research excursions as well—not many people would be willing to spend an afternoon wandering around an abandoned insane asylum.

I want to thank my in-laws, Nancy and Glenn Crosby, who attended nearly every speaking engagement and book signing I gave—their love and encouragement have been steadfast.

Liz Crosby, Meg Crosby, and Scott Crosby have given me their unwavering support as well. And Glenn Crosby, a neurosurgeon, was kind enough to read through some of the technical sections of the book. I am lucky to have them not only as family, but also as friends.

And, as always, I am so grateful for my parents, Tom and Betsy Caldwell, and my sister, Lindsey, who have given me their unconditional love and support in this as in everything.

For their imagination, intelligence, and energy, I thank my two daughters, Morgen Caroline and Keller Elizabeth. Morgen, who writes two or three short stories a day, could have finished this book in half the time it took me, and Keller, though not yet steady with a pen, is a prolific verbal storyteller. They inspire me each and every day.

And to my husband, Andrew—thank you. Without your support, time, patience, insight, and encouragement, none of this would be possible.

NOTES

EPIGRAPHS

The Oliver Sacks quote in the author's note is from the foreword to *Awakenings* (1990 edition). Constantin von Economo's introductory quote is from his book *Encephalitis Lethargica*, published in English in 1931. The second introductory quote, by Oliver Sacks, is again from *Awakenings*, appendices.

PROLOGUE: INSIDE

Most of the information in the prologue came directly from family accounts. Unfortunately, my grandmother had already passed away by the time I was writing the book, so I never had the opportunity to ask her any specific questions. However, her experience was a fairly typical one, even a positive one in that she survived with minimal damage and was never institutionalized.

CASE HISTORY ONE

Chapter 1: An Epidemic Begins

Details about the battle of Verdun were taken from several sources: Britannica; a PBS documentary entitled "The Great War and Shaping of the 20th Century," aired in 1996; Sir Alistair Horne's book *The Price of Glory* (1994); Jon E. Lewis's *The Mammoth Book of Eyewitness World War I* (2003); and John Keegan's *The First World War* (2000).

The quotations about the French dispatches came from Lewis's book, pages 194–95. The estimate about the number of casualties at Verdun—700,000—was consistent in Britannica, the PBS documentary, and Horne's book. The strange image of horses wearing gas masks during chemical battles came from a photograph on www.sciencemuseum.com.

Stanley Weintraub's *Silent Night* (2001) provided the details about the 1914 Christmastime truce on pages 43–44 and 55.

Physical observations about the battle, horse-drawn ambulances, and field hospitals came from photographs found in my sources or additional online image collections.

The majority of information about Jean-René Cruchet's original cases came from Constantin von Economo's account in *Encephalitis Lethargica* (1931) and an article he published, "Cruchet's Encephalomyelite and Epidemic Encephalitis Lethargica," *Lancet* (1929). Von Economo maintained that Cruchet's diagnosis of encephalitis, even if it appeared a few days before his own, was too broad and generalized to recognize the outbreak as one single disease, while von Economo himself was the one to identify the disrupted sleep patterns as the common link.

The last sentence in chapter 1 includes the phrase, "By definition, that made it an epidemic." Most definitions of an epidemic refer to an outbreak among a large number of people at the same time. However, the term can also be applied to one or two isolated patients when new cases of a certain disease occur in a given human population during a given period. In fact, the reemergence of an infectious disease or a single urban case of a disease like yellow fever can be considered an epidemic because of the potential danger.

Chapter 2: Constantin von Economo

Biographical information about von Economo was published in a book by his wife and Julius Wagner-Jauregg, *Baron Constantin von Economo* (1937), which included details like the types of books von Economo kept on his nightstand, as well as all quotes attributed to him. Interestingly, S. E. Jelliffe wrote the foreword to the book. Facts about Wagner-Jauregg's Nobel Prize win in 1927 can be found on www.nobelprize.org. His most famous studies dealt with using fever—in this case malaria—to treat mental illnesses like neurosyphilis.

Von Economo's quote about quitting flying for medicine came from his wife's book, page 26.

Von Economo, in *Encephalitis Lethargica*, recorded the full account of how he came to identify the disease, the history of "nona," and his findings. Tilney, Jelliffe, and Sacks all referred to the nona epidemic in their own writings.

I also consulted an article by J. M. Pearce, "Baron Constantin von Economo and Encephalitis Lethargica," *Journal of Neurology, Neurosurgery and Psychiatry* (1996).

Chapter 3: The London Outbreak

Descriptions of the battle near the river Somme can be found in the PBS series *The Great War* (1996) and John Keegan's *The First World War* (2000).

The account of a British private, named Henry Tandey, encountering a wounded Adolf Hitler on the battlefield was taken from "The Man Who Didn't Shoot Hitler," by Gary Smailes, www.victoriacross.wordpress.com (March 12, 2007) and "Victoria Cross Heroes: Private Tandey Spared the Life of a German Soldier," *Birmingham Post* (July 31, 2004). The story also appears in other World War I accounts, but has never been proven factual. However, according to the accounts, not only Tandey, but also Hitler recounted the tale to contemporaries. Hitler purchased a painting of the battle, pointing out Tandey's image in the portrait and explaining that the British private had spared his life.

All information about Hitler's illnesses and wounds during World War I—particularly the theory that he may have had encephalitis lethargica—came from three articles: P. P. Mortimer's "Was Encephalitis Lethargica a Post-influenzal or Some Other Phenomenon?" *Epidemiology and Infection* 137 (2009); J. Walters's "Hitler's Encephalitis: A Footnote to History," *Journal of Operational Psychology* 6 (1975); and A. Lieberman's "Adolf Hitler Had Postencephalitic Parkinsonism," *Parkinsonism and Related Disorders* 2 (1996). Additional details about Hitler's health can be found in Robert G. L. Waite's *The Psychopathic God* (1977).

Theories about food poisoning and botulism when encephalitis lethargica first hit London were reported in the London *Times* on January 7, 1919, and January 23, 1919.

Peretz Lavie, in *The Enchanted World of Sleep* (1996), suggested that the tales of Sleeping Beauty and Rip van Winkle might have been based on epidemics of encephalitis lethargica, pages 239–40. Oliver Sacks, in *Awakenings*, also referred to his patients as Rip van Winkles and Sleeping Beauties, page 65. There is no documented evidence that Edgar Allan Poe's short stories were based on encephalitis lethargica. However, as Lavie points out, no other disease causes deathlike sleep and eventual waking. After all, these patients were not in a coma; they were sleeping and could usually be roused for short periods of time. The great epidemic did inspire some new stories during the 1920s as well, including one reviewed in the *New York Times* called "A Romance of Two Centuries" about a captured man who is given a sleeping sickness germ and does not wake until 2025.

Details about the London outbreak primarily came from A. J. Hall's *Epidemic Encephalitis* (1924) and the *Memorandum on Encephalitis Lethargica* (1924), held in the History of Medicine Division, National Library of Medicine. Hall also published articles on the subject. I used four of his articles in researching the London outbreaks:

"Epidemic Encephalitis," *British Medical Journal* (October 26, 1918), "The Lumeian Lectures on Encephalitis Lethargica," *Lancet* 1 (1923), "The Mental Sequelae of Epidemic Encephalitis in Children," *British Medical Journal* 1 (1925), and "Note on an Epidemic of Toxic Ophthalmoplegia associated with Acute Asthenia and Other Nervous Manifestations," *Lancet* (April 20, 1918).

Facts about epidemiology were based on Charles Rosenberg's *Explaining Epidemics and Other Studies in the History of Medicine* (1992). Hippocrates' definition of epidemiology is from "The Nature of Man," published as part of *Hippocratic Writings* (1984). Notable examples of epidemiology included John Snow's mapping of cholera from Steven Johnson's *The Ghost Map* (2007); Robert Koch's biography on www.nobelprize.org; Walter Reed's work with typhoid and yellow fever in my book *The American Plague* (2006); and Judith W. Leavitt's *Typhoid Mary* (1996).

CASE HISTORY TWO

Chapter 4: New York City

Historical information about New York City came from a number of sources: Ric Burns and James Sanders's *New York* (2003), as well as Burns's "New York: A Documentary Film"; Nathan Silver's *Lost New York* (2000); David Stravitz's *New York, Empire City 1920–1945* (2004); Kenneth Jackson and David Dunbar's *Empire City* (2002); and E. B. White's *Here Is New York* (2005 reprint).

In addition to those sources, I used some original material like Elizabeth McCausland's book *New York in the Thirties* (1939), featuring photos by Berenice Abbott; Mary Black's *Old New York in Early Photographs* (1973); Laura Spencer Porter's guidebook *New York, the Giant City* (1939); Tony Sarg's *Up and Down New York* (1926); and William J. Showalter's "New York: The Metropolis of Mankind," *National Geographic* (July 1918). Often, it was through studying those books or photos from the 1920s that I found details like store names, carts selling particular goods, the chaotic streets, or other minor details.

To immerse myself in details from the time period, I also reread some classic novels about New York culture, like *The Great Gatsby* and *The Fountainhead*. They are filled with finer points like the fact that they used "ashcans," not trashcans. Removing gloves and hats was the first thing anyone did upon entering a home. Milk came delivered in wooden crates. A number of textural details also came from *New York Times* articles. Although I read through the archives of most newspapers from the time period, like the *New York Evening Post*, the *Herald*, and the *Tribune* at the New York Public Library, I was able to access *New York Times* archives online, and so those archives make up the bulk of newspaper accounts.

It was Laura Spencer Porter's guidebook, *New York, the Giant City*, that pointed out New York grew faster than any of the world's largest cities.

The quotes from E. B. White came from the 2005 reprint of his original *Here Is New York*, page 40.

The historian who wrote New York was "the one place in the world where the hand of man shaped the environment as much as the hand of God" was Kenneth T. Jackson in an essay in Burns and Sanders's book, page 307.

Words used to describe New York during that time period: colossal (Paul Bourget in *Outre-Mer: Impressions of America*, 1895); magnetic (reporter Henry Tyrrell in *New York World*, 1923); astonishing (New York reporter Alva Johnson); feverish (E. B. White); glittering (F. Scott Fitzgerald); an imperial city (Calvin Coolidge in 1925); the Niagara of American Life (Joseph W. Showalter's "New York: The Metropolis of Mankind," *National Geographic* July 1918); and poetry itself (Ezra Pound).

Laura Spencer Porter's original 1939 guidebook provided information about the "City of Cities" and various neighborhoods, pages 24–25.

New York's position after World War I was described in Burns and Sanders's book as the financial center of world, page 315; publishing capital of United States, page 300; and fashion center, page 308.

Descriptions of and details about New York in the winter of 1917–18, the coldest on record, came from *New York Times* articles: "13 below Zero on Coldest Day" (December 31, 1917), "Drastic Action Increases City's Supply of Coal" (January 4, 1918), "New Cold Wave Balks Railroads" (January 21, 1918), "Heat Ban Will Be Stricter" (January 28, 1918), "Workless Mondays" (January 30, 1918), "Coldest Winter on Every Count Record Breaker" (February 10, 1918), and "Coldest Winter Weather Bureau Has Ever Known" (February 10, 1918). Some of the descriptive details also came from personal observation during research trips I took to New York, including one during a February snow. Descriptions of New York street scenes during that time period came from old photographs and newspaper advertisements.

Details about the way people dressed, "facial forestation," wide-brimmed hats, icy hems, mass-produced garments, the "bob," and the need for "bobby pins" all came from Mark Sullivan's *Our Times* (1937), pages 390–411.

Information about New York pollution and soft-coal grease appeared often in newspaper accounts. The "impassable yellow fog" was reported in the *New York Times* (January 19, 1926).

Biographical information about Frederick Tilney came from a variety of places. The Henry Alsop Riley Papers and the Walter Timme Papers held in Columbia University's Augustus C. Long Health Sciences Library provided some of the personal details. Tilney's obituary in *Time* magazine also provided essential pieces of information. The quote about the brain of modern man came from Tilney's acclaimed book *The Brain* (1928).

Details like the "crown breezes," fever boxes, and other early neurological practices came from Lawrence Pool's *Neurological Institute of New York, 1909–1974* (1975). Pool was a neurosurgeon at the institute during some of its pivotal years. His book can be found in Columbia's Health Sciences Library or the New York Academy of Medicine.

The story of Phineas Gage is well documented. I based my description on two sources: www.brainconnection.com and P. Ratiu and I. F. Talos's article "The Tale of Phineas Gage, Digitally Remastered," *New England Journal of Medicine* 351 (December 2, 2004).

Additional information about neurology came from Arthur Link's *Fifty Years of American Neurology* (1998).

All material about early radiology came from two sources: Joel Howell's *Technology in the Hospital* (1995) and Bettyann Kevles's *Naked to the Bone* (1997). The quote about the X-ray threatening the two holiest sanctums came from Kevles's book, page 27, as did material about Clarence Dally and his death, page 47. Details like the coin-operated X-ray machines came from Howell's book, page 137, and Foot-O-Scopes came from Kevles's book, page 80.

Aside from the main biographical facts about Tilney, the story about young Fred Tilney as a medical student came from Roy Chapman Andrews's autobiography, *Under a Lucky Star* (2007). Andrews later became the director of the American Museum of Natural History.

The quote about future generations calling the early neurologists "pioneers" came from Pool's history, page 2. Tilney's article "Professors of Government" appeared in the *Washington Post* (April 8, 1930).

Further details about the original Neurological Institute came from Pool's book, as well as quoted descriptions of Tilney, pages 17 and 124. Tilney's obituary published in *Time* magazine named him as "the country's greatest specialist on brain function" (May 6, 1940).

The quotation about medical progress came from Sullivan's book, page 60.

"Ruth's" case was recorded as "Case V" in Tilney and Howe's book *Epidemic Encephalitis*, published in 1920. Ruth was not her real name. All details about Ruth came from Tilney's case history: her age, medical records, symptoms, treatment, and death. She first contracted the illness on December 19, 1918. She died eight weeks later. In his book, Tilney also described other symptoms, including the fact that patients could hear what was going on around them, as well as the "fixed expression" and waxy, "death-like" face. And, finally, Tilney recounted how physicians had stood around her bedside and remarked that she would probably not recover. It was then that Tilney saw tears running down the girl's face.

In chapter 6 of his book, Tilney describes the case of the four-year-old boy who may have been one of the first cases seen in New York City—if not the first. The boy's symptoms began in September 1916, which would have preceded the earliest recorded cases in 1918 in New York. Tilney's quote about the "unusual and prolonged somnolence" is from his book, page 46.

Tilney's typed medical form was found in the Matheson Files at the Augustus C. Long Health Sciences Library in the Columbia University medical complex.

The "effigy on tomb" description, although referenced in Tilney's book, originally appeared in an article, "Epidemic Encephalitis," by A. J. Hall in the *British Medical Journal* (October 26, 1918).

According to Ruth's chart in Tilney's book, her pulse had raced higher than 170, and her temperature was almost 107 degrees when she died.

Chapter 6: The Neurologist

The account of Tilney's medical relationship with Helen Keller was taken from three main sources: Dorothy Herrmann's *Helen Keller* (1999), pages 241–44; Merlin Donald's *A Mind So Rare* (2002), pages 234–37; and Emily Davis's article "Helen Keller Shows Future of Brain," *Science News-Letter* 14, no. 387 (September 8, 1928), pages 142–44.

Reference to John Daniel, the gorilla, came from Tilney's *The Brain*, page viii.

Details about Tilney treating *New York Times* owner Adolph Ochs came from Susan Tift and Alex Jones's *The Trust* (2000), pages 108 and 140.

Material about Georgia O'Keeffe's stepdaughter as a patient of Tilney's appeared in Benita Eisler's *O'Keeffe and Stieglitz* (1992), pages 325–27.

Chapter 7: The Medical Investigators

Tilney's February 4, 1919, "Address of the Retiring President: The Opportunity of American Neurology," appeared in the *Minutes and Proceedings of the New York Neurological Society*. Tilney's vision for New York as the neurological center of the world was also discussed in depth in Kenton Kroker's article "Epidemic Encephalitis and American Neurology, 1919–1940," *Bulletin of the History of Medicine* 78, no. 1 (Spring 2004). Kroker's article provides an excellent analysis of New York neurology and the role encephalitis lethargica played in its development.

Mention of sleeping sickness first appeared in the *New York Times* on March 11, 1919, "Dr. Copeland Tells of the Appearance of a Rare Disease."

The quote asserting that "no occurrence in the field of neurology" was as illuminating as encephalitis was published in Robert Sheehan's article "Epidemic

Encephalitis," *Psychiatric Quarterly* 2 (March 1928). He also read the paper at the quarterly conference at Manhattan State Hospital on Ward's Island, New York.

The quote about the coordination of neurological research coalescing around epidemic encephalitis came from Kroker's article.

To immerse myself in New York's public health system from the 1890s through the 1930s, I relied on several sources: David Rosner's *Hives of Sickness* (1995); Sandra Opdycke's *No One Was Turned Away* (1999); the *Annual Report of the Department of Health of the City of New York for the Year 1914*; the U.S. Department of Health and Human Services' *Public Health in New York City in the Late Nineteenth Century*, based on an exhibit at the National Institutes of Health (1990); and newspaper accounts.

The most astounding details like the seventeen thousand horses, mules, and cattle removed from the city, along with various animals such as an alligator and two camels, appeared in a *New York Times* article (April 15, 1923), but I used the reports of the *New York Sanitary Code, Sections 87–90*. The *New York Times* also reported on the public bathhouses and laundry trucks (November 22, 1925), bans on public funerals (February 14, 1921), open-air classrooms for TB patients (September 25, 1921), and the ill effects of smoking in public (February 2, 1922).

The majority of historical information about the health department was taken from Rosner's book. Biographical material about William Park came from Morris Schaeffer's "William H. Park (1863–1939)," *American Journal of Public Health* 17, no. 11 (November 1985).

The *New York Times* (April 29, 1918) followed the story of Tammany's failed attempt at control of the health department during the flu pandemic.

Statistics about Ellis Island appeared in Rosner's book, page 70.

Typhoid Mary's tale is well-known in medical history, but it was also told in Burns and Sanders's book, page 275, and Rosner's book, page 109. I also used Manon Parry's "Sara Josephine Baker," *American Journal of Public Health* 96 (2006) and Judith W. Leavitt's *Typhoid Mary* (1996). Additional information about the precarious relationship between the health department and immigrants was taken from Rosner's book, pages 37 and 73.

Details about the 1916 polio epidemic in New York City can be found in an online exhibit for the National Museum of American History; Jeffrey Kluger's *Splendid Solution* (2006); and David Oshinsky's *Polio* (2006). After centuries of battling diseases associated with filth, it was the first time physicians began to question the relationship between cleanliness and *declining* immunities. Medical studies noted that people living on farms had healthier immune systems than those who lived in cities. The reason: farm families were exposed to a wider variety of microbes.

Sources for the 1918 flu are discussed in the notes for chapter 24, "Gray Matter."

The quote in the *New York Times* about New York's health department as a model of municipal health appeared on April 21, 1918.

Health Commissioner Royal Copeland's campaign to improve public health was closely followed by the *New York Times* in articles published on March 31, 1919, and July 12, 1919. The *New York Times* reported the average life expectancy in the 1920s as fifty years. That is consistent with Sullivan's book, which listed life expectancy in 1920 as fifty-five years.

Rosner's book, page 155, remarked upon New York's health system being recognized worldwide.

Sigmund Freud's quote was published in his *Mass Psychology*, page 121 (2004 edition).

CASE HISTORY THREE

Chapter 8: Adam

Like Ruth, "Adam's" name is fictional. However, all other details are factual, from the train ride home to the symptoms to the quotes. The account of Adam's illness was published more than once, each offering a little more or different personal details. S. E. Jelliffe published the case in a series of articles, "Postencephalitic Respiratory Disorders: Review of the Syndromy, Case Reports and Discussion," *Journal of Nervous and Mental Disease* 63 (July–December 1926). In those articles Jelliffe also included some details from Charles Burr, the original doctor to see Adam, who published his case study, "Sequelae of Epidemic Encephalitis without Any Preceding Acute Illness," *Archives of Neurology and Psychiatry* 14 (1925). Jelliffe disagreed with Burr's original diagnosis and believed Adam's spring case of the flu was in fact an acute case of encephalitis lethargica.

Chapter 9: Smith Ely Jelliffe

Most biographical material about Jelliffe came from John Burnham's *Jelliffe* (1983). The childhood recollections of Jelliffe were featured in Burnham's book, pages 9–17.

Some additional information about Jelliffe's involvement in the nona epidemic and his quotes about misdiagnosing women or "bumble puppy" came from an article Jelliffe published, "Nervous and Mental Disturbances of Influenza," *New York Medical Journal* 108 (November 2, 1918). That journal, as well as many of the other, older medical periodicals, is held at the New York Academy of Medicine.

Jelliffe's quote about his wife's involvement in science came from Burnham's book, page 16.

Further details about Jelliffe were taken from *Psychoanalytic Pioneers*, edited by Franz Alexander et al. (1996), the reference to Jelliffe's first published article about plants in Prospect Park, page 225, and Jelliffe as a fierce opponent in medical debates, page 228.

For some of the more sensational details about Jelliffe, like the trial of Henry K. Thaw, I followed articles in the *New York Times* (March 1907–08). Jelliffe's testimony in the Albert Fish trial was reported in the *New York Times* (March 21, 1935).

The quote about early Freudian theories came from Burnham's book, page 70.

The quotation about the human organism and the whole evolutionary product was published by Jelliffe in his "Postencephalitic Respiratory Disorders" article.

The description of Jelliffe as one who gazes backward, takes the path to the present, and analyzes what is to come came from his eulogy.

Burnham's book, page 87, describes the death of Jelliffe's first wife and his marriage to his second wife, Bee Dobson. And the colorful stories about Jelliffe as a father came from Burnham's book, pages 148–50, including the account of his friendship with Lionel Barrymore. Observations from Mabel Dodge also came from Burnham's book, pages 144–45. Included are some of the other famous patients of Jelliffe's like Betty Compton and playwright Eugene O'Neill.

Chapter 10: The Alienist

Descriptions of Central Park were taken primarily from Roy Rosenzweig and Elizabeth Blackmar's *The Park and the People* (1992). Additional details like roasted corn in the handcarts came from various *New York Times* articles or studying historical photographs.

I found details about the various train routes and elevated trains from old photographs and historical maps from www.nycsubway.org, as well as in Clay McShane's *Down the Asphalt Path* (1994). For typical foods sold or created during the 1920s, I again turned to online searches or looking at old menus, often on display in historic hotels.

The fact that 32,000 speakeasies existed during the 1920s came from Burns and Sanders's book, page 318, but was also reported in the National Archives, which estimated anywhere between 30,000 to 100,000 speakeasies in New York City.

The article about the seminar on alcohol use and Jelliffe's participation appeared in the *New York Times* (May 16, 1919). And Copeland's remarks about cocaine use and draft dodgers appeared in the *New York Times* (April 15, 1919).

Information about the ways automobiles changed America, as well as the photograph of Fifth Avenue in 1900 and again in the 1920s, came from Sullivan's book, page 375, and McShane's book. The quote about disappearing surface travel and second-story sidewalks came from Sullivan's book, page 368, but was originally quoted in the *New York World* (May 10, 1903).

The series of acts passed by Congress during that short time span include: Selective Service Act of 1917 (the draft), passed by Congress on May 18, 1917; the Espionage Act, passed June 15, 1917; and the Sedition Act (an amendment to the Espionage Act) passed on May 16, 1918.

Details about what became known as the First Red Scare came from *New York Times* coverage on April 30, May 1, May 2, and June 3 in 1919. In addition to those attacks, a post office in New York discovered sixteen packages with bombs, and in 1920 came the famed Wall Street explosion outside J. P. Morgan's bank.

Other worries of the time period, like automobile deaths, came from Sullivan's book, page 380, as did parental concerns of the age, pages 430–31.

The description of Adam was based on photos published in Jelliffe's article "Postencephalitic Respiratory Disorders." Details about Adam's visits to Jelliffe's office came from the same article, and again, all quotes are authentic.

The story of Dora Mintz, the New York woman who slept for over one hundred days before waking to the sound of a violinist playing Schubert's "Serenade," appeared in a *New York Times* article on January 21, 1920, and a London *Times* article on January 22, 1920.

All material about Dr. Oliver Sacks and the patients at the Beth Abraham Hospital came directly from *Awakenings* or the interview I conducted with him in December 2007. The quote about the inextricable link between psychiatry and neurology came from that interview.

Jelliffe's letter to Freud was reprinted (along with most of his correspondence to Freud or Jung) in Burnham's book, page 215.

Chapter 11: Only the Beginning

The *New York Times* provided coverage and details about Charles Lindbergh's historic parade on June 13, 1927. Famous photographs, like the one that appeared in Burns and Sanders's book, page 354, depicted the snowstorm of confetti and stock exchange tickers.

As I said in the book, there is no proof Adam attended the parade, but his meetings that week with Jelliffe did include accounts of attending parties, and it would have been surprising if a young man like Adam did not take part in the celebratory atmosphere of the city that week.

Jelliffe's paper "Psychologic Components in Postencephalitic Oculogyric Crises" was published in the *Archives of Neurology and Psychiatry* 21 (1929).

Descriptions of Jelliffe's street, Fifty-sixth Street, came from personal observation—Jelliffe's brownstone is still standing—and details about his office were depicted in photographs published in Burnham's book. Information about Cornelius Vanderbilt's recently demolished house came from Silver's *Lost New York*, page 110.

Jelliffe's account of Adam's trance and quote appeared in his article "Psychological Components in Postencephalitic Oculogyric Crises."

The quote from the New York neurologist Dr. Charles Burr appeared in the article, "The Mental Disorders of Childhood," *American Journal of Psychiatry* 82 (July 1925).

The estimated nine thousand articles published on encephalitis lethargica during the epidemic or immediately following it came from Oliver Sacks and Joel Vilensky's article, "Waking to a New Flu Threat," *New York Times* (November 16, 2005).

The quote about "altered dispositions" came from a London *Times* article (October 13, 1922).

CASE HISTORY FOUR

Chapter 12: Jessie

The F. Scott Fitzgerald quote about the city in summer like overripened fruit came from *The Great Gatsby*: "I love New York on summer afternoons when everyone's away. There's something very sensuous about it—overripe, as if all sorts of funny fruits were going to fall into your hands."

Descriptions of Penn Station were based on photographs published in Silver's *Lost New York*, pages 32–38.

I found newsstand magazine covers by searching out vintage cover art online. For example, American author Ellis Parker Butler's website, www.ellisparkerbutler.com, shows examples of his magazine cover stories, including the June 1925 cover for *Ladies' Home Journal*. Norman Rockwell's *Saturday Evening Post* covers are now collector's items for sale. *Good Housekeeping* and *Field and Stream* June 1925 covers were found through similar methods. Information about the first few copies of the *New Yorker*, which hit the stands in February 1925, was taken from its website: www.newyorker.com.

Descriptions of Glen Cove and the Morgan home there came from Ron Chernow's excellent book *The House of Morgan* (1990). Photos of the estate are included in Chernow's book. All personal and biographical information about the Morgans came from Chernow's book, including their marriage, page 64; Jack's early aspirations to be a doctor, page 64; Jack's philandering father, page 98; Morgan's relationship with

Jessie and the fear of his father's loveless fate, page 171; and the 1915 shooting in the Morgans' Glen Cove home, pages 177 and 193.

Accounts of Jessie's illness were covered briefly in Chernow's book, page 267, but most details were taken from *New York Times* coverage during the summer of 1925: June 18, 19, 20, 21, 22, 23, 25, 27, and 29 and July 3, 6, and 27. Each report indicated Jessie's health was improving.

Jack Morgan's quote about Jessie's slow recovery appeared in Chernow's book, page 268.

Just as coverage of Jessie's illness was waning in the constant news cycle, the *New York Times* had to report her death in the August 15, 1925, issue. Details about her funeral and the posted notice were published in the *Times* (August 16, 1925). Details about the servants spread along the coast to keep reporters at bay came from the *Herald-Tribune* (June 18, 1925).

The account of Charles Rothschild's case of encephalitis lethargica and eventual suicide came from Niall Ferguson's *The House of Rothschild,* vol. 2, *The World's Banker: 1849–1999* (2000), pages 451–52. Rothschild's obituary was also published in the London *Times* (October 15, 1923).

In fact, suicide was fairly common in adults like Rothschild suffering from chronic symptoms. The *New York Times* reported that one woman committed suicide and left a note that said, "You will find my body in the lower lake." And a student at Yale committed suicide by turning on the gas in his room.

Chernow's book, page 268, offered further details about Jack Morgan's reaction to Jessie's death, tending to her garden, and keeping her bedroom exactly the same.

The Henry Riley Papers at Columbia's Health Sciences Library described Tilney approaching Morgan for the $200,000 donation, which funded an entire floor of the Neurological Institute. In fact, Tilney would be credited with raising over one-half the contributions that helped build the new institute.

Chapter 13: 1925

The progress laws mentioned in this chapter include the Indian Citizenship Act of 1924; the Eighteenth Amendment, enacting Prohibition, passed on January 16, 1920; and the Nineteenth Amendment, allowing women the vote passed on August 18, 1920.

In 1949, Dr. Alexander Kennedy of Durham University, at a meeting of the American Association for the Advancement of Science, suggested that "the great American gangster era of the late Twenties and early Thirties was linked with an epidemic of encephalitis which lasted from 1918 to 1926." And Dary Matera's book *John Dillinger* (2004) mentioned Kennedy's theory on page 400, although there is no link between Dillinger himself and encephalitis lethargica.

Biographical information about presidents Warren G. Harding and Calvin Coolidge can be found on the White House's website, www.whitehouse.gov. Biographers of Harding have written that his death may have been caused by stroke, pneumonia, exhaustion, food poisoning, or poisoning. It was the controversial—though popular—biography by Gaston Means entitled *The Strange Death of President Harding* (1930) that suggested Harding's own wife could have been to blame. Means's accusation was unfounded and is disregarded by most historians.

The darker side of the progress laws included a surge in anti-immigrant sentiment. During the 1920s, the Ku Klux Klan had as many as five million members across the United States. The Immigration Act or National Origins Quota Act was passed in 1921, and the Johnson-Reed Act was passed in 1924.

In trying to verify whether or not New York City truly did have more Italians than Rome, Germans than Berlin, or Irish than Dublin, I found references to those statistics in a *New York Times* article (June 4, 1908) and William Joseph Showalter's article "New York: The Metropolis of Mankind," *National Geographic* (July 1918).

Though both suffrage and Prohibition had been political issues for decades, Sidney M. Milkis and Jerome M. Mileur's *The New Deal and the Triumph of Liberalism* (2002) asserted that those two progress laws may have been aimed at the immigrant lifestyle, pages 316–17.

Famed advocate Margaret Sanger in *The Birth Control Review* 1, no. 1 (February 1917) outlined her ideas for what would become the American Birth Control League, a precursor to Planned Parenthood.

The American Neurological Association's *Semi-Centennial Anniversary Volume of the American Neurological Association, 1875–1924* was edited by both Tilney and Jelliffe.

The tragic story of Jelliffe's son and his accidental death came from Burnham's book, page 95, as well as newspaper accounts. The letter Jelliffe wrote to Freud about the tragedy was published in Burnham's book, page 217.

I found details about Tilney's social life and trip to see Eleanora Duse in the *New York Times* (October 30, 1923). The "Tilney Memorial" published in *Time* magazine on May 6, 1940, carried the incredible story of Tilney's stroke and recovery.

Tilney's quote came from *The Brain* (1928), page 776.

Chapter 14: A Two-Headed Beast

Jelliffe's quote about the diversified types of mental illness caused by encephalitis lethargica came from his article "Nervous and Mental Disturbances of Influenza."

The average age of Parkinson's during the encephalitis lethargica epidemic—thirty-six—was taken from Melvin Yahr's article "Parkinsonism before and since the Epidemic of Encephalitis Lethargica," *Archives of Neurology* 9 (September 1963).

The statistic about the number of children who showed psychological changes after a case of encephalitis lethargica came from T. R. Hill's article "Problem of Juvenile Behaviour Disorders in Chronic Epidemic Encephalitis," *Journal of Neurology and Psychopathology* 9 (1928).

The "fatal wound healed, the second beast was coming forth" refers to the seven-headed beast from the sea followed by the beast from the Earth in Revelation 13.

CASE HISTORY FIVE

Chapter 15: Madness

The opening quote was taken from E. D. Bond and G. E. Partridge's article "Postencephalitic Behavior Disorders in Boys and Their Management in Hospital," *American Journal of Psychiatry* 6, no. 25 (1925).

The London *Times* published an article (June 6, 1924) describing the crimes and violent behavior appearing in the wake of the encephalitis lethargica epidemic.

The examples of insanity described by von Economo came from his book *Encephalitis Lethargica* (1931).

Further examples of violent behavior, like suicide and assault attempts, came from Bond and Partridge's article as well as H. D. MacPhail's article "Mental Disorder from Encephalitis Lethargica," *Journal of Mental Science* 68 (1922).

A child's drawing depicting flames bursting from a head and the violent story of a murdered child appeared in the *American Journal of Psychiatry* 10, no. 5 (March 1931).

The tragic description by one patient, "It's so sad to be like me . . . ," came from Elizabeth Bixler's "The Nurse and Neurological Problems," *American Journal of Nursing* 35, no. 5 (May 1935).

The London *Times* reported on the Mental Deficiency Act and its amendment to include postencephalitic patients in articles published on December 4, 1926; January 14, 1926; and June 29, 1927.

Numerous other articles are listed in the bibliography that were written by physicians during the 1920s treating patients—most often children—showing mental problems following a case of encephalitis lethargica.

Tilney's quote appeared in a chapter of a book discussing the opening of a new children's unit, though Tilney's remark refers specifically to the Children's Unit at Kings Park: "Children's Unit Opened at Rockland State Hospital," Notes and Comments, *Mental Hygiene* (National Committee for Mental Hygiene, 1937), page 157.

Much of the physical description of Kings Park came from personal observation during a visit to the asylum in February 2008. At the time, I also interviewed Steve Weber from the Kings Park Museum and was given a tour of the grounds. Weber provided me with historic photos of the asylum during the 1920s, as well as descriptions of how day-to-day life was conducted on a farm asylum.

I also relied upon Leo Polaski's *The Farm Colonies* (2003).

An article by August E. Witzel, "Epidemic Encephalitis, Sequelae and the Psycho-neuroses," *State Hospital Quarterly* 10, no. 3 (May 1925), outlined the differences between psychopathic children and postencephalitic children. The *State Hospital Quarterly*, now out of print, is held at the New York Academy of Medicine.

Mary Boyle's *Schizophrenia* (1990), pages 70–75, suggests that many 1920s neurologists may have been confusing schizophrenia with encephalitis lethargica.

I based my research on mental illness on two main sources: Roy Porter's *Madness* (2002) and Edward Shorter's *A History of Psychiatry* (1997). I also read David Rothman's *The Discovery of the Asylum* (2005), which dealt with early nineteenth-century development of institutionalized care; Richard Bentall's *Madness Explained* (2003); Robert Whitaker's *Mad in America* (2002), which addressed some of the horrific treatments of the mentally ill throughout the last century; and E. Fuller Torrey and Judy Miller's *The Invisible Plague* (2001).

The early history of madness came from Roy Porter's *Madness*, pages 10–11. Bethlehem's transition into "Bedlam" was mentioned in both Porter's book, page 71, and Shorter's book, page 4. The quote about "confinement" in early colonial times in the United States came from Shorter's book, page 7.

Information about mental illness during the age of Enlightenment, the reclassification of a number of mental illnesses, and the original hospitals and "asylums" came from Shorter's book, pages 2, 48–49, and 33–34.

Details about Dorothea Dix appeared in Shorter's book, pages 3–4. Additional biographical information came from Francis Tiffany's *Life of Dorothea Dix* (1891) and "Dorothea Dix and Franklin Pierce," a program on NPR with an article online at www.npr.org.

Incidentally, Dix's idea was not a new one. By far, the most revolutionary approach to psychiatric care is a town in Belgium known as Geel, or Gheel. The town hospital started placing mentally ill patients with families in the village or in the countryside as early as the thirteenth century. Other patients, who lived in the hospital full-time, were allowed to have jobs in the town and walk freely through the village before returning to the hospital at night. At its peak, in 1938, nearly four thousand patients were placed with families in Geel; even today, five hundred patients live with Geel families. It is a community effort and an incredibly successful one at that—the hospital pays a small amount to the families who house the patients, and the town in general supports and helps them. The freedom of a "normal" life with responsibility and work proves far more healing than any hospital environment.

The quote referring to the defeated reformers in asylum care came from Shorter's book, page 46. Likewise, the neurologist who complained of three hundred patients per doctor—famed physician Adolf Meyer—appeared in Shorter's book, pages 46–47. Adolf Meyer became one of America's undisputed leaders in brain study and psychiatry.

Most descriptions of the children's program at Kings Park came from H. A. Robeson's article "The Children's Unit at Kings Park State Hospital," *State Hospital Quarterly* 10, no. 4 (August 1925). The article described how the children were treated, the programs they participated in, the schedule they adhered to, and the attempt at normalcy. All quotes about the children's unit or the children who lived there are attributed to the same article. C. E. Gibbs's "Behavior Disorders in Chronic Epidemic Encephalitis," *American Journal of Psychiatry* 86, no. 4 (January 1930), also provided details, as well as statistics about the number of children who lived there during the 1920s.

The quote about losing the original child forever appeared in A. J. Hall's book *Epidemic Encephalitis* (1924).

Chapter 16: Rosie

Descriptions of the New York Academy of Medicine are based on personal observation. I made several trips there in the course of researching this book. In addition to the main reading room, the NYAM has a beautiful library of old medical texts. Much of the building is unchanged since it was opened in 1926, so the elevator still has a cage door and an elevator operator, and the auditorium where Tilney and Jelliffe sat is still relatively the same.

Information about both Tilney's and Jelliffe's involvement in the New York Neurological Society, which held its meetings at the NYAM, came from the *Minutes of the New York Neurological Society* from 1919 (when both Tilney and Jelliffe were on the officer ballot), as well as the program from S. P. Goodhart's presentation in 1931. Interestingly, Goodhart also gave a lecture that year on the importance of keeping neurology and psychiatry as one entity. The minutes from those meetings are held in the Rare Book Room at the NYAM.

Rosie's story (her name has been changed for this book) appeared in full in S. P. Goodhart and N. Savitsky's "Self-Mutilation in Chronic Encephalitis," *American Journal of the Medical Sciences* 185 (January–June 1933), pages 674–83.

The majority of details about Rosie's case came from Goodhart and Savitsky's original article. However, her case was also covered by Joel A. Vilensky, Paul Foley, and Sid Gilman in their article "Children and Encephalitis Lethargica," *Pediatric Neurology* 37, no. 2 (2007). Vilensky, Foley, and Gilman are three of only a handful of contemporary physicians still actively studying the 1920s epidemic of encephalitis lethargica. Dr. Vilensky was nice enough to send me documentary footage about the disease and to answer various questions I had while researching this book.

The history of self-mutilation addressed in this chapter came from a variety of sources. NPR had a program entitled "The History and Mentality of Self-Mutilation" (June 10, 2005), which told the story of the "needle girls." Goodhart himself referred to Saint Lucia, Oedipus, and biblical verses dealing with self-mutilation in his article about Rosie. Different versions of the Saint Lucia or Saint Lucy story exist. In the Christian tradition, her eyes were gouged out as part of torture. In my book, I used another version, in which Saint Lucia gouged them out herself. The biblical quote about plucking out the right eye comes from Matthew 5:29 (American Standard Version).

Historical details about the original Morrisania Hospital came from Christopher Gray's article "Streetscapes: Morrisania Hospital: A Tidy Relic of the 1920's Looking for a New Use," *New York Times* (July 15, 1990).

The weather report of the storm on the night of July 30 came from the *New York Times* (July 31, 1931). And the actual full moon occurred on July 29, 1931, so it would still have looked "full" one or two days later.

I found the unusual fact that the neighborhood around Morrisania once cultivated mushrooms from an online article at www.forgotten-ny.com. The same question was also addressed in the *New York Times* "FYI" section (July 20, 2003).

Lisa Cartwright's book *Screening the Body* (1995), pages 72–80, included photos from and information about Goodhart's film of Rosie: "Acute Epidemic Encephalitis" (1944).

Chapter 17: The Neurological Institute

The account of the von Economos' voyage on the *Olympic* came from *Baron Constantin von Economo* by Karoline von Economo and Julius Wagner-Jauregg, pages 34–35.

F. Scott Fitzgerald's quote about the "wild promise of all the mystery and the beauty in the world," appeared in *The Great Gatsby*. Ezra Pound's quote about the New York skyline was mentioned in Burns and Sanders's book, page 293, and first appeared in Pound's "Patria Mia," 1912. The Ayn Rand quote came from her classic novel *The Fountainhead* (1943). Rand also wrote, "I would give the greatest sunset in the world for one sight of New York's skyline." And Frank Lloyd Wright wrote about the vertical city in *The Disappearing City* (1932).

Jelliffe's quote about von Economo being "peerless" as a man and a scientist, as well as his analogy comparing great medical advances to the building of a skyscraper, came from his foreword to von Economo's and Wagner-Jauregg's book.

The estimate that New York's tons of rubbish could reach the same heights as the Woolworth Building came from the *New York Times* (May 28, 1928), as did the quote about a New York as clean as Havana.

Statistics about the number of buildings erected between 1928 and 1931 came from Burns and Sanders's book, pages 368–69, and specific details about the Chrysler Building, Bank of Manhattan Co., and Empire State Building came from www.nyc-architecture.com.

For accounts of the stock market crash, I relied on three main sources: John K. Galbraith's *The Great Crash 1929* (1997); "The Crash of 1929," *The American Experience*; and Maury Klein's *Rainbow's End* (2003).

Wall Street's $30 billion loss was mentioned in Burns and Sanders's book, page 376. And Galbraith's book, pages 128–29, reported that, contrary to popular belief, there was *not* a run on the banks. The first suicide from the top of the Statue of Liberty was reported by the *New York Times* (May 14, 1929), several months prior to the crash.

Information about the drought came from "Surviving the Dust Bowl," *The American Experience.*

New York was *arguably* the city hardest hit by the Depression. It may have been a close second to Detroit. The auto industry, like banking, suffered huge losses and declining sales.

The quote from Bernard Sachs appeared in August Wimmer's *Further Studies upon Chronic Epidemic Encephalitis* (1929).

Early history about the Neurological Institute was taken from archives at Columbia's Augustus C. Long Health Sciences Library and Pool's book, which is held in that collection as well as at the NYAM. Additional details about the new building and Tilney's vision came from an article entitled "Unique Three-fold Study of Mental Ills Begins," *New York Times* (September 15, 1929). Tilney's quotes came from that article as well.

CASE HISTORY SIX

Chapter 18: The Matheson Commission

All biographical information about William J. Matheson came from www.keyhistory.org/matheson and his obituary in the *New York Times* (May 16, 1930). Like Jessie Morgan, since Matheson was a public figure whose case of encephalitis lethargica was covered in newspapers, his name was not changed for this book. In using his personal medical files, held in the Charles Loomis Dana Papers (volume 4, case #62) in the Rare Book Room of the New York Academy of Medicine, I used discretion, relying on pertinent medical information and excluding some of the personal family details that physicians of the time period typically included in case studies. Matheson was under Dana's care from 1921 to 1923.

Kroker's article on encephalitis lethargica and American neurology also included material from Dana's medical file and the development of the Matheson Commission. Dana's quote about encephalitis lethargica being exceedingly rare came from Kroker's article.

Chapter 19: Josephine B. Neal

Though discrimination against women in early medical schooling is widely covered, I relied on two sources: Regina Morantz-Sanchez's *Sympathy and Science* (1985) and John Duffy's *From Humors to Medical Science* (1993).

Sullivan's book *Our Times* (1937) describes the transition from farming to city homes and its effect on women, including the details about washing day, ironing day, cleaning day, etc., pages 424–28.

The French social psychologist who compared an intelligent woman to a monstrosity, specifically a two-headed gorilla, was Gustave Le Bon, and he was a student of Paul Broca, the famed craniologist. His statements were taken from Stephen Jay Gould's well-known and popular book *The Mismeasure of Man* (1981). Gould, an anatomist and anthropologist, criticized early brain study as rooted in racism and sexism.

The American writer who said that a woman's name should appear in print but twice was Arthur Wallace Calhoun in his book *A Social History of the American Family from Colonial Times to the Present* (1918), page 326. Unfortunately, he is sometimes accused of making the remark, when in reality, he was quoting a southern gentleman. Calhoun's book highlighted the lack of progress in women's education and social standing.

Morantz-Sanchez wrote about the number of women in law, Ph.D., and medical programs, page 314. Morantz-Sanchez was also the historian who remarked upon Vassar, Bryn Mawr, and Smith sending women to Johns Hopkins for over two decades, page 250. Duffy, too, on pages 294–95, described the statistics: in 1915, 2.6 percent of medical students were women; in 1927, that number had climbed only to 5.4 percent and then dropped again. In fact, according to Duffy, page 288, Harvard Medical School did not admit a female student until 1945.

Duffy, pages 294–95, refers to the unofficial quota systems that prevented minorities from entering medical schools in large numbers, but Duffy also suggests that the attitudes professors held about women students played an even larger part in discrimination.

Marie Curie's two Nobel Prize wins can be found on "Multiple Nobel Laureates," www.nobelprize.org.

The two issues women lobbied for in public health were the United States Children's Bureau, established in 1912, and the Sheppard-Towner Maternity and Infancy Protection Act of 1921 (Morantz-Sanchez, page 296). In fact, one of the pioneers of public health, a nurse named Lillian Wald, had started the movement for home health care as early as 1893—actually going into the tenements and slums to care for ill people and help teach them about child care and sanitation.

The high number of deaths during delivery and births in a hospital vs. a midwife came from Morantz-Sanchez, page 298, as well as Duffy, pages 286–87. Public health as "the woman's branch" was taken from Morantz-Sanchez's book, page 301.

The definitions of meningitis and encephalitis came from a medical dictionary.

Jonas Salk announced his vaccine on April 12, 1955. However, Neal's work with a polio vaccine at the bacteriology lab of New York's health department was reported in the *New York Times* (July 7, 1934). Discussion of testing the polio vaccine can also be found in Lawrence Altman's *Who Goes First?* (1987), pages 126–28.

Descriptions of New York streets came from historic photos, including one that depicts piles of refuse on street corners.

Details about the New York Academy of Medicine came from personal observations during many visits there.

The quote about encephalitis lethargica as one of the "most imperative problems the science of medicine has to solve," appeared in a *New York Times* article, "Light on Encephalitis Gained by New Survey" (November 17, 1929).

Chapter 20: Vaccine Trials

The history of the CDC can be found on its website, www.cdc.gov. Originally called the Communicable Disease Center, the name was later changed to the Centers for Disease Control and Prevention. The center was founded in 1946 as an organization primarily for entomologists and engineers studying mosquito-borne diseases. Some of the more famous investigations include Legionnaires' disease (1976), Marburg virus (1967), Ebola (1976), and HIV (1983–84).

Information about the Matheson Commission and its vaccine work came almost entirely from the Matheson Files held at the Augustus C. Long Health Sciences Library at Columbia University and the three reports published by the Matheson Commission: *Epidemic Encephalitis: Etiology, Epidemiology, Treatment* (1929, 1932, 1939). All three reports are available at the New York Academy of Medicine and Columbia's Health Sciences Library. Neal also published a book entitled *Encephalitis*

(Grune and Stratton, 1942) that provided a number of statistics about the 1920s epidemic of encephalitis, particularly in New York.

I also consulted E. D. Louis's article "Vaccines to Treat Encephalitis Lethargica: Human Experiments at the Neurological Institute of New York, 1929–1940," in *Archives of Neurology* 59, no. 9 (September 2002).

Information about Rosenow's work came from Kroker's article in the *Bulletin of the History of Medicine*, as well as an article published by Neal, "The Present Status of the Etiology of Epidemic Encephalitis," *Journal of the American Medical Association* 91 (1928). Kroker's article went on to explain the fact that Neal's position as a woman made her somewhat of an outsider.

Material about William Park's lab and bacteriologist Anna Williams came from Morantz-Sanchez's book, page 160. Additional biographical information about Williams was taken from Rosner's book, *Hives of Sickness*, page 171. In fact, Williams was the bacteriologist who isolated the diphtheria bacillus—the research surrounding diphtheria and its antitoxin helped put the bacteriology labs of the New York health department on the map of great medical research centers.

The quote about Park's "harem" came from Rosner's book, page 171, and originally appeared in Anna Williams's autobiography.

Details about Matheson's heart attack aboard his yacht came from his *New York Times* obituary.

Constantin von Economo's three Nobel Prize nominations can be found on www.nobelprize.org.

Early information about neurosurgery came from Lawrence Pool's book about the Neurological Institute. Pool, who was among the first generation of neurosurgeons, told the colorful stories about early surgery at Bellevue, the patient who smoked a cigarette during surgery, and the "initiation" nurses had for new surgeons.

Chapter 21: Sylvia

Though her name has been changed to protect her privacy, I found "Sylvia's" case history in the Melvin D. Yahr Personal Papers and Manuscripts at Columbia's Augustus C. Long Health Sciences Library. Yahr turned over a number of the Matheson Commission case files.

The letter dated September 24, 1942, came directly from her file. In fact, one of the reasons I chose her case study was because of the ongoing correspondence between Sylvia and her doctors.

Estimates about the number of women who served in the military during WWI—as many as forty-nine thousand served in uniform—came from Brian P. Mitchell's *Women in the Military* (1998), page 3. And descriptions of the conditions came from the firsthand accounts in various World War I histories like Horne's book *The Price of Glory*.

Rarely is there a consensus when it comes to diagnosing—retroactively—an illness of the past, especially when it involves someone as important as a U.S. president. Not surprisingly, Woodrow Wilson's illness at the close of the war has been debated among historians. Edwin Weinstein's *Woodrow Wilson* (1981), pages 336–40, suggested that a series of strokes earlier in Wilson's life, as well as a bout with influenza, followed by an encephalitis virus (as many physicians considered encephalitis lethargica to be) led to neurological complications and an altered disposition during the Paris peace negotiations. Alexander George's *Presidential Personality and Performance* (1998), pages 69–80, on the other hand, disputed those findings, finding little evidence of previous strokes and believing instead that Wilson was behaving as he always did—stubbornly. John M. Barry, in his book *The Great Influenza* (2004), expanding upon the theory of Alfred Crosby, believed Wilson acted as anyone might while suffering from a terrible case of the flu. The quotes I used about aides reporting the "queer changes" in Wilson's personality after the sickness came from Barry's book, pages 385–88. What is certain is that Wilson suffered a severe case of the flu, and whatever the cause of his mindset, his policies changed. As Barry wrote, "Historians with virtual unanimity agree that the harshness toward Germany of the Paris peace treaty helped create the economic hardship, nationalistic reaction, and political chaos that fostered the rise of Adolf Hitler."

Details about the film footage Tilney took of patients came from Lisa Cartwright's book *Screening the Body*. Tilney referred to it as "bradykinetic," or a "slow movement analysis."

Information about encephalitis lethargica as a possible inspiration for the zombie films of the 1920s and '30s came from Gary Don Rhodes's *White Zombie: Anatomy of a Horror Film* (1997). On page 44 of his book, Rhodes mentions encephalitis lethargica, as well as catatonia, as inspirations for zombie films. Many films from that era, like *The Cabinet of Dr. Caligari*, also described sleeping sicknesses and sleepwalking.

The quote from one of Neal's patients who said, "I am become mentally subnormal," was published in Neal's book *Encephalitis* (1942), page 344.

All details came from her file and the correspondence between Sylvia and her doctors, as well as correspondence between the doctors and Sylvia's family, including the final letter dated March 20, 1945.

Chapter 22: I Have Seen the Future

Information about Mayor Jimmy Walker in the 1930s came from Edward Robb Ellis's book *The Epic of New York City* (2004), page 526. Further information about

Walker was taken from Burns and Sanders's book, pages 416–18. Details about La Guardia's election came from Burns and Sanders's book, pages 420–23; Silver's book, pages 258–59; and Ellis's book, pages 551–52. The quote about the condition of New York when La Guardia took office came from Ellis's book, page 551.

Paul Starr's *The Social Transformation of American Medicine* (1982), pages 270–71, pointed out that all the civic improvements in New York during this time ultimately hurt public health, which received much less funding.

Ellis's book, page 553, also described the building of the Rockefeller Center and the fact that John D. Rockefeller, Jr., was able to employ thousands of people during the Depression.

White's well-known quote about New York reaching the highest point in the sky at the lowest point of the Depression appears on page 30 in the reprint of his book.

Details about how Central Park changed in the 1930s came from Rosenzweig and Blackmar's *The Park and the People*, pages 441–54, and Silver's book, pages 59 and 244. Descriptions of "Hooverville" came from Silver's book, page 258.

For descriptions of the 1939 World's Fair, I studied old photos available through a number of historical collections, as well as relying upon Silver's book, page 259, and Andrew Wood's *New York's 1939–1940 World's Fair* (2004). The *New York Times* also published details about various exhibits at the fair in the article "World's Fair of '39 Revisited" (June 20, 1980).

The full letter that Neal wrote to the patients who took part in the vaccine trials is held in the Matheson archives at Columbia's Health Sciences Library.

The pleas to keep the program open for the sake of the patients was made by Matheson Committee member Hubert S. Howe in a letter dated June 14, 1939. Kroker's quote was taken from his article "Epidemic Encephalitis and American Neurology, 1919–1940."

The letter requesting cab fare for Neal (dated October 14, 1940) was found in the Matheson files and archives. Neal retired one year later.

Details about Neal's life and death came from her obituary, *New York Times* (March 20, 1955).

Details about Tilney's death and final years came from his obituaries in the *New York Times* and the *New York Post*, both August 8, 1938.

The final division between neurology and psychiatry was described by Jack Pressman in his book *Last Resort* (1998), page 39.

Details about Jelliffe's later years came from Burnham's biography of him, page 100. Charles Burlingame published an article, "The Jelliffe Library," *Science* (April 18, 1941), about the collection of books Jelliffe left. Information about psychosurgery came from Pressman's book, as well as Shorter's, pages 227–29. Shorter also described the famous lobotomy cases, like Tennessee Williams's sister, and I verified information with Tennessee Williams biographies and Kennedy biographical material.

The use of neuroleptics was described by Shorter, page 254, as well as in Whitaker's book, pages 151–53. Whitaker remarked upon the similarity between the "chemical lobotomies" and the symptoms of encephalitis lethargica.

The quote from Alfred Crosby's book *America's Forgotten Pandemic* appears in the afterword, page 325.

The death of William O'Dwyer's wife during his term as mayor of New York was published in the *New York Times* (October 13, 1946), and her death was reported as cardiac failure resulting from complications, including "Parkinson's disease condition and post-encephalitis lethargica."

Neal's book *Encephalitis*, page 326, quoted Jelliffe: "In the monumental strides made by psychiatry during the past ten years no single advance has approached in importance that made by the study of epidemic encephalitis." Neal went on to write, "This disease has thereby assumed a role of some significance in providing a means of bridging the known gaps between neurology and psychiatry."

The neurologist who remarked that no other disease affected so large a portion of its victims for so long a time was C. E. Gibbs. And the 1986 article calling encephalitis lethargica a "disease of momentous importance for three decades" was written by Christopher Ward, "Encephalitis Lethargica and the Development of Neuropsychiatry," *Psychiatric Clinicians of North America* 9, no. 2 (1986).

Von Economo's quote came from his book *Encephalitis Lethargica* (1931).

Oliver Sacks's quote was taken from *Awakenings*, the foreword to the 1990 edition. As one patient said to Sacks, "Tell our story, or it will never be known" (BBC documentary and interview with Sacks).

CASE HISTORY SEVEN

Chapter 23: Philip

Philip Leather's story was part of the BBC documentary "Medical Mysteries: The Forgotten Plague," which aired in England in 2004.

His physician G. A. Auden (the poet's father) was one of the close observers of the epidemic in England. He wrote an article, "Encephalitis Lethargica: Its Psychological Implications," *Journal of Mental Science* 71 (October 1925), in which he described not only mental defects caused by the illness, but also the reverse—bright thinking, creativity, and intelligence.

Chapter 24: Gray Matter

Anyone interested in the 1918 influenza pandemic is lucky enough to have three excellent choices on the subject: Alfred Crosby's *America's Forgotten Pandemic* (1990); Gina Kolata's *Flu* (2001); and John M. Barry's *The Great Influenza* (2004). They are not only factual and thorough accounts of the pandemic but also engrossing narratives.

Information about Dr. John Oxford's story was taken from Kolata's book and the BBC documentary "Medical Mysteries: The Forgotten Plague," aired in July 2004. Kolata's book covers his groundbreaking work with the 1918 influenza virus, and the BBC documentary details Oxford's work involving encephalitis lethargica. The BBC also included an article on its website, "Mystery of the Forgotten Plague," on July 27, 2004 (www.news.bbc.co.uk).

Kolata interviewed Oxford in the course of researching her book, and the bulk of the story takes place in chapter 10. Her quote about encephalitis lethargica's connection to the 1918 flu is on page 294. Information about the experience of Western Samoa and American Samoa was taken from pages 294–95.

Dr. Jeffrey Taubenberger was another physician searching for a connection between the flu and encephalitis lethargica. His group (A. H. Reid et al.) worked in the United States and published the article "Experimenting on the Past," *Journal of Neuropathology and Experimental Neurology* 60, no. 7 (July 2001).

All information about Sophie Cameron came from the website dedicated to her: www.thesophiecamerontrust.org.uk. Sophie became ill in 1999 and spent nine months in the hospital. Her bout with encephalitis lethargica left her severely brain damaged and physically handicapped. Sophie died in 2006, at the age of twenty-four.

The BBC documentary also told the story of Becky, a modern-day patient who baffled British doctors with her strange symptoms. They tested her for everything from mumps to hepatitis, measles to arboviruses. Once they treated her with steroids, she began a slow recovery, but it took months for her to recover fully. Becky spent the next two years learning to walk and talk again.

I found the story about the high school student from Texas in an online publication by the University of Texas Health Science Center at Houston (July 10, 2007), under the title "Jarrod's Story."

Dr. Joel Vilensky's article "The 'Spanish' Influenza Epidemic of 1918 and Encephalitis Lethargica," available on www.thesophiecamerontrust.org.uk, provided details about the relationship between the flu and encephalitis lethargica. The book published in Britain that suggests a definite link between influenza and encephalitis lethargica is Niall Johnson's *Influenza Pandemic*, published in 2006.

Dr. Russell Dale's early work with encephalitis lethargica was covered in the

BBC documentary on the subject. His original studies linking the disease to strep were conducted with Dr. Andrew Church in London. I also interviewed Dr. Dale in the course of my research, and all details about his work since he left London for the University of Sydney were taken from that interview.

Further information was taken from articles published by Dale and his colleagues: "Encephalitis Lethargica Syndrome," *Brain: A Journal of Neurology* 127 (January 2004) and "Contemporary Encephalitis Lethargica Presenting with Agitated Catatonia, Stereotypy, and Dystonia-Parkinsonism," *Movement Disorders* (November 2007).

Frederick Tilney also refers to von Economo's research, as well as that of von Wiesner, in isolating gram-positive diplostreptococcus in a case in 1917. Tilney said that the tissue reaction to this infection was inflammation (Tilney and Howe, *Epidemic Encephalitis*, 1920).

The article referring to the streptococci from St. Elizabeth's Hospital was "The Germ of Sleeping Sickness," *Science* 78, no. 2018 (September 1933).

Chapter 25: Past or Prologue?

The 2008 study about pandemic flu and the number of deaths resulting from pneumonia came from John Brundage and G. Dennis Shanks's "Deaths from Bacterial Pneumonia during 1918–19 Influenza Pandemic," Centers for Disease Control, *Emerging Infectious Diseases*, vol. 14, no. 8 (August 2008). I also consulted a publication by the CDC, *A Commentary on the JAMA Study's Interpretation of the Influenza Experience in New York City and Chicago, 1918–19* (2008), and an article by David M. Morens and colleagues: "Predominant Role of Bacterial Pneumonia as a Cause of Death in Pandemic Influenza," *Journal of Infectious Diseases* (2008).

The War Information Office published *The Medical Clinics of North America*, U.S. Army, vol. 2, no. 2 (September 1918), which charted the number of strep and staph infections rampant among soldiers.

The number of scarlet fever deaths came from New York health department records: Arthur Cosby's *New Code of Ordinances of the City of New York*, adopted June 20, 1916, with all amendments to January 1, 1922.

The reference to "satellite infections" appeared in A. J. Hall's book *Epidemic Encephalitis* (1924).

As suggested in the chapter, there is no medical evidence linking encephalitis lethargica to the multiple infectious diseases of the 1920s, nor to the vaccines and antitoxins. Those are my own opinions and theories about a disease that has provided so few answers. However, it is not necessarily a new idea. An article/chapter published in 1932 by Dr. Earl B. McKinley in a publication for the Association for Research in

Nervous and Mental Disease asked the same question: "But is epidemic encephalitis always due to one cause? Or is this disease caused by a variety of agents?"

Estimates about the number of people afflicted by encephalitis lethargica during the pandemic vary greatly. At the time of the pandemic, most physicians believed that cases were *under*diagnosed because the symptoms were unclear, and an acute infection was not always obvious. Five million was the general estimate for decades. However, more recent work conducted by Joel Vilensky and his colleagues suggests the number of patients could be much lower because physicians of the time period used the disease as a catchall for a number of conditions.

It was A. J. Hall, in his book *Epidemic Encephalitis* (1924), who described an isolated case as early as 1903.

Neal's book *Encephalitis* reported on the timeline of the epidemic, with earliest dates for the outbreak in Bucharest in 1915, France in 1916–17, and Vienna in 1917.

Epilogue: Virginia and the Forgotten Epidemic

W. H. Auden's poem is entitled "Nothing to Save." As he was the son of Dr. George Auden, who treated a number of British children with encephalitis lethargica, I was intrigued by the fact that his father's work may have influenced some of his own poetry.

The quote from Sacks's *Awakenings* appeared in the prologue, page 22, of the 1990 edition.

My grandmother, Virginia Thompson Brownlee, passed away in 1998 in Dallas, Texas. Her death was not related to her case of encephalitis lethargica or any complicating factors.

BIBLIOGRAPHY

HISTORICAL COLLECTIONS

AUGUSTUS C. LONG HEALTH SCIENCES LIBRARY, COLUMBIA UNIVERSITY

Archives of Neurological Institute

Matheson Commission Files

Melvin D. Yahr Personal Papers and Manuscripts

Walter Timme Papers

Henry Alsop Riley Papers

KINGS PARK HERITAGE MUSEUM COLLECTION

MUNICIPAL ARCHIVES OF THE CITY OF NEW YORK

Annual Report of the Department of Health of New York City for the Year 1914.

New York City Sanitary Code, Sections 87, 89, 90.

NATIONAL LIBRARY OF MEDICINE, History of Medicine Division

Memorandum on Encephalitis Lethargica (Great Britain, Ministry of Health, 1924).

Health, Disease and Integration: An Essay based on a Study of Certain Aspects of Encephalitis Lethargica (Henry Pratt Newsholme, 1929).

NEW YORK ACADEMY OF MEDICINE

Epidemic Encephalitis: Etiology, Epidemiology, Treatment, by the Matheson Commission (first report, 1929; second report, 1932; third report, 1939).

Charles Loomis Dana Papers, volume 4, case #62.

State Hospital Quarterly, volumes 6–11, 1920–1930.

Minutes and Proceedings of the New York Neurological Society, 1918–1924, 1925–1934.

Committee on Public Health, Minutes, March 17, 1919.

NEW YORK PUBLIC LIBRARY, Humanities and Social Sciences Library

New York City Newspapers Collection

BOOKS

Acute Epidemic Encephalitis: An Investigation by the Association for Research in Nervous and Mental Diseases: Report of the Papers and Discussions at the Meeting of the Association, New York City, December 28th and 29th, 1920. New York: Association for Research in Nervous and Mental Disease, 1921.

Alexander, Franz, Samuel Eisenstein, Martin Grotjahn, eds. *Psychoanalytic Pioneers.* New York: Basic Books, 1996.

Allemann, Albert. *The Medical Interpreter: The Interpretation and Translation of the World's Practical Medicine and Surgery.* Chicago: Pepin and Pridgen, 1921.

Altman, Lawrence. *Who Goes First? The Story of Self-Experimentation in Medicine.* 1986. Berkeley: University of California Press, 1998.

Andrews, Roy Chapman. *Under a Lucky Star: A Lifetime of Adventure.* Speath Press, 2007.

Barry, John M. *The Great Influenza: The Epic Story of the Deadliest Plague in History.* New York: Viking, 2004.

Bentall, Richard P. *Madness Explained: Psychosis and Human Nature.* 2003. New York: Penguin Global, 2005.

Black, Mary. *Old New York in Early Photographs.* New York: Dover, 1973.

Blake, Angela M. *How New York Became American, 1890–1924.* Baltimore: Johns Hopkins University Press, 2006.

Booss, John, and Margaret M. Esiri. *Viral Encephalitis in Humans.* Washington, DC: American Society for Microbiology, 2003.

Boyle, Mary. *Schizophrenia: A Scientific Delusion?* London: Routledge, 1990.

Burnham, John C. *Jelliffe: American Psychoanalyst and Physician.* Chicago: University of Chicago Press, 1983.

Burns, Ric, and James Sanders. *New York: An Illustrated History.* New York: Knopf, 2003.

Calhoun, Arthur Wallace. *A Social History of the American Family from Colonial Times to the Present.* Cleveland: The Arthur H. Clark Company, 1918.

Cartwright, Lisa. *Screening the Body: Tracing Medicine's Visual Culture.* Minneapolis: University of Minnesota Press, 1995.

Chase, Josephine. *New York at School: A Description of the Activities and Administration of the Public Schools of the City of New York.* Ann Arbor: Scholarly Publishing Office, University of Michigan Library, 2005.

Chernow, Ron. *The House of Morgan: An American Banking Dynasty and the Rise of Modern Finance.* New York: Grove Press, 1990.

Copland, James, John Darwall, and John Conolly. *The London Medical Repository and Review.* London: Thomas and George Underwood, 1826.

Cosby, Arthur. *New Code of Ordinances of the City of New York: The Sanitary Code, the Building Code and Park Regulations*. 1916.

Crosby, Alfred W. *America's Forgotten Pandemic: The Influenza of 1918*. New York: Cambridge University Press, 1990.

Crosby, Molly Caldwell. *The American Plague*. New York: Berkley Books, 2006.

Donald, Merlin. *A Mind So Rare: The Evolution of Human Consciousness*. W. W. Norton and Co., 2002.

Duffy, John. *From Humors to Medical Science: A History of American Medicine*. Urbana, IL: University of Illinois Press, 1993.

Eisler, Benita. *O'Keeffe and Stieglitz: An American Romance*. New York: Penguin, 1992.

Ellis, Edward Robb. *The Epic of New York City: A Narrative History*. 1966. New York: Basic Books, 2004.

Elsberg, Charles. *The Story of a Hospital: The Neurological Institute of New York, 1909–1938*. New York and London: Paul Hoeber, 1944.

Ferguson, Niall. *The House of Rothschild*, vol. 2, *The World's Banker 1849–1999*. 1998. New York: Penguin, 1999.

Finger, Stanley. *Origins of Neuroscience: A History of Explorations into Brain Function*. New York: Oxford University Press, 1994.

Foley, Paul B. *Beans, Roots, and Leaves: A History of the Chemical Therapy of Parkinsonism*. Marburg, Germany: Tectum Verlag, 2003.

Freud, Sigmund. *Mass Psychology*. 1921. New York: Penguin, 2004.

Galbraith, John Kenneth. *The Great Crash 1929*. 1954. New York: Mariner Books, 1997.

George, Alexander L., and Juliette L. George. *Presidential Personality and Performance*. Boulder, CO: Westview Press, 1998.

Goldberg, Elkhonon. *The Executive Brain: Frontal Lobes and the Civilized Mind*. New York: Oxford University Press, 2001.

Gould, Stephen Jay. *The Mismeasure of Man*. New York: W. W. Norton and Co., 1981.

Hall, A. J. *Epidemic Encephalitis (Encephalitis Lethargica)*. London: Simpkon, Marshall, Hamilton, Kent and Co., 1924.

Halpern, Sydney. *Lesser Harms: The Morality of Risk in Medical Research*. Chicago: University of Chicago Press, 2004.

Hays, J. N. *The Burdens of Disease: Epidemics and Human Response in Western History*. New Brunswick, NJ: Rutgers University Press, 1998.

Herrmann, Dorothy. *Helen Keller: A Life*. 1998. Chicago: University of Chicago Press, 1999.

Homberger, Eric. *The Historical Atlas of New York City: A Visual Celebration of 400 Years of New York City's History*. 1994. New York: Henry Holt and Co., 1998.

Hood, Clifton. *722 Miles: The Building of the Subways and How They Transformed New York*. New York: Simon and Schuster, 1993.

Horne, Sir Alistair. *The Price of Glory: Verdun 1916*. New York: Penguin, 1994.

Howell, Joel D. *Technology in the Hospital: Transforming Patient Care in the Early Twentieth Century*. 1995. Baltimore: Johns Hopkins University Press, 1996.

Jackson, Kenneth T., and David S. Dunbar. *Empire City: New York Through the Centuries*. 2002. New York: Columbia University Press, 2005.

Jelliffe, Smith Ely. *Postencephalitic Respiratory Disorders*. New York: Nervous and Mental Disease Publishing Company, 1927.

Johnson, Steven. *The Ghost Map: The Story of London's Most Terrifying Epidemic—and How It Changed Science, Cities, and the Modern World*. 2006. New York: Riverhead, 2007.

Keegan, John. *The First World War*. 1999. New York: Vintage, 2000.

Kevles, Bettyann H. *Naked to the Bone: Medical Imaging in the Twentieth Century*. Reading, MA: Helix Books, 1997.

Klein, Maury. *Rainbow's End: The Crash of 1929*. New York: Oxford University Press, 2003.

Kluger, Jeffrey. *Splendid Solution*. New York: Putnam, 2005.

Kolata, Gina. *Flu: The Story of the Great Influenza Pandemic of 1918 and the Search for the Virus That Caused It*. 1999. New York: Touchstone, 2001.

Kroker, Kenton. *The Sleep of Others and the Transformations of Sleep Research*. Toronto: University of Toronto Press, 2007.

Lash, Joseph, and Trude Lash. *Helen and Teacher: The Story of Helen Keller and Anne Sullivan Macy*. 1980. New York: Da Capo Press, 1997.

Lavie, Peretz. *The Enchanted World of Sleep*. New Haven, CT: Yale University Press, 1996.

Lay, Maxwell G. *Ways of the World: A History of the World's Roads and of the Vehicles That Used Them*. New Brunswick, NJ: Rutgers University Press, 1992.

Leavitt, Judith Walzer. *Typhoid Mary: Captive to the Public's Health*. Boston: Beacon Press, 1996.

Lewis, Jon E., ed. *The Mammoth Book of Eyewitness World War I*. New York: Carroll and Graf, 2003.

Lewis, Nolan D. C. "Smith Ely Jelliffe 1866–1945: Psychosomatic Medicine in America," in *Psychoanalytic Pioneers*, edited by F. Alexander, S. Eisenstein, and M. Grotjahn. New York: Basic Books, 1966.

Link, Arthur S., ed. *Fifty Years of American Neurology: An Historical Perspective: A Semicentennial Essay by Smith Ely Jelliffe*. 1924. Winston-Salem, NC: Stratford Books, 1998.

Lishman, W. A., ed. *Organic Psychiatry*. Oxford: Blackwell, 1987.

Lloyd, G. E. R., ed. and J. Chadwick, trans. *Hippocratic Writings*. New York: Penguin Classics, 1984.

Matera, Dary. *John Dillinger: The Life and Death of America's First Celebrity Criminal*. New York: Carroll and Graf, 2004.

McCausland, Elizabeth, and Berenice Abbott. *New York in the Thirties*. New York: Dover, 1939.

McKinley, Earl R. "Etiology of Epidemic Encephalitis," chapter 9 in *Infections of the Central Nervous System: An Investigation of the Most Recent Advances*. Association for Research in Nervous and Mental Disease. Baltimore: Williams and Wilkinson, 1932.

McShane, Clay. *Down the Asphalt Path: The Automobile and the American City*. New York: Columbia University Press, 1994.

Meara, Frank Sherman. *The Treatment of Acute Infectious Diseases*. New York: Macmillan, 1921.

Milkis, Sidney M., and Jerome M. Mileur, eds. *The New Deal and the Triumph of Liberalism*. Amherst, MA: University of Massachusetts Press, 2002.

Mitchell, Brian P. *Women in the Military: Flirting with Disaster*. Washington, DC: Regnery, 1998.

Morantz-Sanchez, Regina. *Sympathy and Science: Women Physicians in American Medicine*. 1985. Chapel Hill: University of North Carolina Press, 2000.

Neal, Josephine B. *Encephalitis*. New York: Grune and Stratton, 1942.

Opdycke, Sandra. *No One Was Turned Away: The Role of Public Hospitals in New York City since 1900*. New York: Oxford University Press, 1999.

Oshinsky, David. *Polio: An American Story*. New York: Oxford University Press, 2006.

Peretz, Lavie. *The Enchanted World of Sleep*. New Haven, CT: Yale University Press, 1993.

Pinto, Vivian de Sola, and F. W. Roberts, eds. *The Complete Poems of D. H. Lawrence*. New York: Penguin, 1971.

Polaski, Leo. *The Farm Colonies: Caring for New York's Mentally Ill in Long Island's State Hospitals*. New York: Kings Park Heritage Museum, 2003.

Pool, J. Lawrence. *Neurological Institute of New York, 1909–1974: With Personal Anecdotes*. Lakeville, CT: Pocket Knife Press, 1975.

Porter, Laura Spencer. *New York, the Giant City: An Introduction to New York*. New York: Treasure Tower, 1939.

Porter, Roy. *Madness: A Brief History*. New York: Oxford University Press, 2003.

Pressman, Jack D. *Last Resort: Psychosurgery and the Limits of Medicine*. New York: Cambridge University Press, 1998.

Public Health in New York City in the Late Nineteenth Century. Exhibit September–December 1990. Washington, DC: U.S. Department of Health and Human Services, Public Health Service, National Institutes of Health, 1990.

Rayburn, Charles R. *Epidemic Encephalitis*. Providence, RI: Johnson, 1929.

Rhodes, Gary Don. *White Zombie: Anatomy of a Horror Film*. McFarland, 1997.

Rosenberg, Charles. *Explaining Epidemics and Other Studies in the History of Medicine*. New York: Cambridge University Press, 1992.

Rosenzweig, Roy, and Elizabeth Blackmar. *The Park and the People: A History of Central Park*. New York: Cornell University Press, 1992.

Rosner, David. *Hives of Sickness: Public Health and Epidemics in New York City*. New Brunswick, NJ: Rutgers University Press, 1995.

Rothman, David. *The Discovery of the Asylum: Social Order and Disorder in the New Republic*. New York: Aldine Transaction, 2002.

Rowland, Lewis P. *The Legacy of Tracy J. Putnam and H. Houston Merritt*. New York: Oxford University Press, 2009.

Sacks, Oliver. *Awakenings*. 1973. New York: Harper Perennial, 1990.

Sarg, Tony. *Up and Down New York*. New York: Universe, 1926.

Shorter, Edward. *A History of Psychiatry: From the Era of the Asylum to the Age of Prozac*. New York: John Wiley and Sons, 1997.

Silver, Nathan. *Lost New York*. 1967. New York: Houghton Mifflin, 2000.

Starr, Paul. *The Social Transformation of American Medicine*. New York: Basic Books, 1982.

Stravitz, David. *New York, Empire City 1920–1945*. New York: Harry N. Abrams, 2004.

Sullivan, Mark. *Our Times: The United States, 1900–1925*. New York: Charles Scribner's Sons, 1937.

Tiffany, Francis. *The Life of Dorothea Lynde Dix*. Boston and New York: Houghton, Mifflin and Company, 1891.

Tift, Susan E., and Alex S. Jones. *The Trust: The Private and Powerful Family Behind the New York Times*. 1999. New York: Back Bay Books, 2000.

Tilney, Frederick. *The Brain: From Ape to Man*. New York: Paul B. Hoeber, 1928.

———, and Hubert Howe. *Epidemic Encephalitis*. New York: Paul B. Hoeber, 1920.

———, and S. E. Jelliffe, eds. "Fifty Years of American Neurology: Fragments of an Historical Retrospect," in *Semi-Centennial Anniversary Volume of the American Neurological Association, 1875–1924*. New York: American Neurological Association, 1924.

Timme, Walter. *Acute Epidemic Encephalitis (Lethargic Encephalitis): An Investigation by the Association for Research in Nervous and Mental Diseases*. New York: Paul B. Hoeber, 1920.

Torrey, E. Fuller, and Judy Miller. *The Invisible Plague: The Rise of Mental Illness from 1750 to the Present*. 2001. New Brunswick, NJ: Rutgers University Press, 2007.

Von Economo, C. *Encephalitis Lethargica: Its Sequelae and Treatment*. Trans. K. O. Newman. London: Oxford University Press, 1931.

Von Economo, Karoline and Julius Wagner-Jauregg. *Baron Constantin von Economo: His Life and Work*. Chicago: University of Chicago Press, 1937.

Waite, Robert George Leeson. *The Psychopathic God: Adolf Hitler.* New York: Basic Books, 1977.

Weinstein, Edwin. *Woodrow Wilson: A Medical and Psychological Biography.* Princeton, NJ: Princeton University Press, 1981.

Weintraub, Stanley. *Silent Night: The Story of the World War I Christmas Truce.* New York: Free Press, 2001.

Whitaker, Robert. *Mad in America: Bad Science, Bad Medicine, and the Enduring Mistreatment of the Mentally Ill.* Cambridge, MA: Perseus Books, 2002.

White, E. B. *Here Is New York.* 1949. New York: Little Book Room, 2005.

Wimmer, S. A. *Chronic Epidemic Encephalitis.* London: Heinemann, 1924.

———. *Further Studies upon Chronic Epidemic Encephalitis.* London: Heinemann, 1929.

Wood, Andrew F. *New York's 1939–1940 World's Fair.* Charleston, SC: Arcadia, 2004.

JOURNAL ARTICLES

Abrahamson, Isador. "Neurological Complications of Influenza (Proceedings of NYNS on January 7, 1919). *Journal of Nervous and Mental Disorders* 49 (1919).

Annual Report of Emma Pendleton Bradley Home. East Providence, RI (1933).

Anonymous. "Children's Unit Opened at Rockland State Hospital." *Mental Hygiene* 21 (1937).

———. "Disabilities." *Lancet* (December 4, 1948).

———. "Encephalitis: A Clinical Study, Book Review." *Journal of American Medical Association*, 119, no. 6 (June 6, 1942).

———. "Epidemic Encephalitis in England." *Science* 66, no. 1701 (August 5, 1927).

———. "Epidemics." *Time* (July 23, 1928).

———. "George Augustus Auden." *Lancet* (May 11, 1957).

———. "The Germ of Sleeping Sickness," *Science* 78, no. 2018 (September 1933).

———. "Notes." *Natural History: Journal of the American Museum of Natural History* 21, no. 2 (1921).

———. "Sections of Psychiatry, Neurology, Epidemiology, and Diseases of Children." *Lancet* (February 7, 1925).

———. "Tilney Memorial." *Time* (May 6, 1940).

Ashwal, Stephen, and Robert Rust. "Child Neurology in the 20th Century." *Pediatric Research* 53 (February 2003).

Auden, G. A. "Encephalitis Lethargica: Its Psychological Implications." *Journal of Mental Science* 71 (October 1925).

Basheer, S. N., L.D. Wadsworth, and B.H. Bjornson. "Anti-basal Ganglia Antibodies and Acute Movement Disorder following Herpes Zoster and Streptococcal Infections." *European Journal of Pediatric Neurology* 11 (2007).

Beverly, B. I., and M. Sherman. "Post-encephalitic Behavior Disturbance without Physical Signs." *American Journal of Diseases in Children* 27 (1924).

Bixler, Elizabeth S. "The Nurse and Neurological Problems." *American Journal of Nursing* 35, no. 5 (May 1935).

Bond, E. D., and Kenneth E. Appel. "The Treatment of Post-encephalitic Children in a Hospital School." Read to the annual meeting of the American Psychiatric Association, Washington, DC, May 5–9, 1930. *American Journal of Psychiatry* 10, no. 5 (March 1931).

———, and G. E. Partridge. "Post-encephalitic Behavior Disorders in Boys and Their Management in a Hospital." *American Journal of Psychiatry* 6, no. 25 (1925).

Breggin, P. R. "Encephalitis Lethargica." *Journal of Neuropsychiatry and Clinical Neurosciences* 7, no. 3 (Summer 1995).

Bromberg, Walter. "Mental States in Chronic Encephalitis." *Psychiatric Quarterly* 4 (October 1930).

Brundage, John, and G. Dennis Shanks. "Deaths from Bacterial Pneumonia during 1918–19 Influenza Pandemic." Centers for Disease Control, *Emerging Infectious Diseases* 14, no. 8 (August 2008).

Burlingame, C. Charles. "The Jelliffe Library." *Science* (April 18, 1941).

Burr, Charles W. "Crime from a Psychiatrist's Point of View." *Journal of the American Institute of Criminal Law and Criminology* 16, no. 4 (February 1926).

———. "The Mental Disorders of Childhood." *American Journal of Psychiatry* 82 (July 1925).

———. "Sequelae of Epidemic Encephalitis without Any Preceding Acute Illness," *Archives of Neurology and Psychiatry*, 14 (1925).

Carlson, Earl R. "Understanding and Guiding the Spastic." *American Journal of Nursing* 39, no. 4 (April 1939).

Casals, J., T. S. Elizan, and M. D. Yahret. "Postencephalitic Parkinsonism: A Review." *Journal of Neural Transmission* 105, no. 6–7 (September 1998).

Centers for Disease Control. *A Commentary on the JAMA Study's Interpretation of the Influenza Experiences in New York City and Chicago, 1918–19* (2008).

Cheyette, S. R., and J. L. Cummings. "Encephalitis Lethargica: Lessons for Contemporary Neuropsychiatry." *Journal of Neuropsychiatry and Clinical Neurosciences* 7, no. 2 (Spring 1995).

Cole, Blanche E. "The Problem of Social Adjustment following Epidemic Encephalitis in Children." *Mental Hygiene* (October 1924).

Cruchet, J. R. "The Relation of Paralysis Agitans to the Parkinsonian Syndrome of Epidemic Encephalitis." *Lancet* 1 (1927).

Cummings, Jeffrey L., Tiffany Chow, and Donna Mastermanet. "Encephalitis Lethargica: Lessons for Neuropsychiatry." *Psychiatric Annals* 31, no. 3 (2001).

Dale, Russell, Andrew J. Church, Robert A. H. Surtees, Andrew J. Lees, Jane E. Adcock, Brian Harding, Brian G. R. Neville, and Gavin Giovannoni. "Encephalitis Lethargica Syndrome: 20 New Cases and Evidence of Basal Ganglia Autoimmunity." *Brain: A Journal of Neurology* 127, part 1 (January 2004).

———, Richard Webster, and Deepak Gill. "Contemporary Encephalitis Lethargica Presenting with Agitated Catatonia, Stereotypy, and Dystonia-Parkinsonism." *Movement Disorders* (November 2007).

Davis, Emily C. "Helen Keller Shows Future of Brain." *The Science News-Letter* 14, no. 387 (September 8, 1928).

Dickman, M. S. "Von Economo's Encephalitis." *Archives of Neurology* 58, no. 1 (October 2001).

Dourmashkin, R. R. "What Caused the 1919–30 Epidemic of Encephalitis Lethargica?" *Journal of the Royal Society of Medicine* 90, no. 9 (September 1997).

Duvoisin, R. C., M. D. Yahr, M. D. Schweitzer, and H. H. Merritt. "Parkinsonism Before and Since the Epidemic and Encephalitis Lethargica." *Archives of Neurology* (September 9, 1963).

Ebaugh, F. G. "Neuropsychiatric Sequelae of Acute Epidemic Encephalitis in Children." *American Journal of Diseased Children* 25, no. 89 (1923).

Fairweather, D. S. "Psychiatric Aspects of the Post-encephalitic Syndrome." *Journal of Mental Science* 93, no. 353 (1947).

Flexner, Simon. "Encephalitis and Poliomyelitis." *Proceedings of the National Academy of Sciences* 6, no. 3 (March 15, 1920). Rockefeller Institute for Medical Research, New York. (Read to the Academy on November 10, 1919.)

Foley, Paul. "Encephalitis Lethargica: A Disease Which Makes Criminals, Eleventh Annual Meeting of the Internaitonal Society for the History of Neurosciences," *Journal of the History of Neurosciences* 15, no. 4 (2006).

Folsom, R. P. "Mental Symptoms of Encephalitis." *Journal of Nervous and Mental Disease* 63, no. 154 (February 1926).

Gibbs, C. E. "Behavior Disorders in Chronic Epidemic Encephalitis." *American Journal of Psychiatry* 86, no. 4 (January 1930).

Goodhart, S. P., and N. Savitsky. "Self-Mutilation in Chronic Encephalitis: Avulsion of Both Eyeballs and Extraction of Teeth." *American Journal of the Medical Sciences* 185 (1933).

Grossman, Morris. "Late Results in Epidemic Encephalitis." *Archives of Neurology and Psychiatry* (May 1921).

Hall, A. J. "Epidemic Encephalitis." *British Medical Journal* (October 26, 1918).

———. "The Lumeian Lectures on Encephalitis Lethargica." *Lancet* 1 (1923).

———. "The Mental Sequelae of Epidemic Encephalitis in Children." *British Medical Journal* 1 (1925).

———. "Note on an Epidemic of Toxic Ophthalmoplegia Associated with Acute Asthenia and Other Nervous Manifestations." *Lancet* (April 20, 1918).

Heiman, H. "Postinfluenzal Encephalitis." *American Journal of Diseases in Children* 6 (1919).

Heyman, M. B. "Psychosis with Epidemic Encephalitis." *State Hospital Quarterly* 6, no. 32 (November 1920).

Hill, T. R. "Problem of Juvenile Behaviour Disorders in Chronic Epidemic Encephalitis." *Journal of Neurology and Psychopathology* 9 (1928).

Hoekstra, Lilly D., and Ruth Shaw Scrivner. "The Care of Encephalitis Patients." *American Journal of Nursing* 34, no. 1 (January 1934).

Hohman, L. B. "Post-encephalitic Behavior Disorders in Children." *Johns Hopkins Hospital Bulletin* 33, no. 372 (October 1922).

Hutchings, Richard H. "Psychiatric Work in Vienna." Presented to the Quarterly Conference at Rochester State Hospital, *The Psychiatric Quarterly*, 3 (September 17, 1929).

Ingram, Madelene. "Encephalitis: Story of a Patient." *American Journal of Nursing* 26, no. 6 (June 1936).

Jang, Haeman, David A. Boltz, Robert G. Webster, and Richard Jay Smeyne. "Viral Parkinsonism." *Biochimica et Biophysica Acta* 1792, 7 (July 2009).

Jelliffe, Smith Ely. "Are Neurology and Psychiatry Separate Entities?" *Journal of Nervous and Mental Disorders* 73 (1931).

———. "Encephalitis Lethargica." *New York Medical Journal* 111 (September 21, 1921).

———. "The Mental Pictures in Schizophrenia and in Epidemic Encephalitis." *American Journal of Psychiatry* 6, no. 3 (January 1927).

———. "Nervous and Mental Disturbances of Influenza." *New York Medical Journal* 108 (November 2, 1918).

———. "Postencephalitic Respiratory Disorders." *Journal of Nervous and Mental Disease* 63 (1926), and continuing through eight issues.

———. "Psychologic Components in Postencephalitic Oculogyric Crises: Contribution to a Genetic Interpretation of Compulsion Phenomena." *Archives of Neurology and Psychiatry* 21 (1929).

———. "Somatic Pathology and Psychopathology at the Encephalitis Crossroad: A Fragment." *Journal of Nervous and Mental Disease* 61 (1925).

Jellinger, K. A. "A Short History of Neurosciences in Austria." *Journal of Neural Transmission* 113, no. 3 (March 2006).

Jenkins, R. L., and L. Ackerson. "The Behavior of Encephalitic Children." *American Journal of Orthopsychiatry* 4 (1934).

Kapadia, F., and S. Grant. "Encephalitis Lethargica." *Psychiatric Quarterly* 16, no. 5 (May 1990).

Kirby, G. H., and T. K. Davis. "Psychiatric Aspects of Epidemic Encephalitis." *Archives of Neurology and Psychiatry* 5 (1921).

Kroker, Kenton. "Epidemic Encephalitis and American Neurology, 1919–1940." *Bulletin of the History of Medicine* 78, no. 1 (Spring 2004).

———. "The First Modern Plague: Epidemic Encephalitis in America 1919–1939." *Transactions and Studies of the College of Physicians of Philadelphia* 24 (December 2002).

Lieberman, A. "Adolf Hitler Had Postencephalitic Parkinsonism." *Parkinsonism and Related Disorders* 2, no. 2 (April 1996).

Louis, E. D. "Vaccines to Treat Encephalitis Lethargica: Human Experiments at the Neurological Institute of New York, 1929–1940." *Archives of Neurology* 59, no. 9 (September 2002).

MacPhail, H. D. "Mental Disorder from Encephalitis Lethargica." *Journal of Mental Science* 68 (1922).

Magee, Kenneth R. "Parkinson's Disease." *American Journal of Nursing* 55, no. 7 (July 1955).

Malzberg, Benjamin. "Age of First Admissions with Encephalitis Lethargica." *Psychiatric Quarterly* , no. 2 (June 1929).

Marshall, R. M. "Mental Aspects of Epidemic Encephalitis." *Journal of Mental Science* 73 (1927).

McCarthy, Tom. "The Coming Wonder? Foresight and Early Concerns about the Automobile." *Environmental History* (January 2001).

McCowan, P. K. "Observations on Some Cases of Post-encephalitic Psychosis." *Lancet* (February 7, 1925).

McGarr, T. E. "Fifty Years of Development in the Case of the Insane in New York State." *Psychiatric Quarterly* (1928).

McKenzie, I. "Discussion of Epidemic Encephalitis: Epidemiological Considerations." *British Medical Journal* (1927).

Morens, David M., Jeffery K. Taubenberger, and Anthony S. Fauci. "Predominant Role of Bacterial Pneumonia as a Cause of Death in Pandemic Influenza: Implications for Pandemic Influenza Preparedness." *Journal of Infectious Diseases* (2008).

Mortimer, P. P. "Was Encephalitis Lethargica a Post-influenzal or Some Other Phenomenon?" *Epidemiology and Infection* 137 (2009).

Neal, Josephine B. "Encephalitis." *American Journal of Nursing* 26, no. 6 (June 1936).

———. "Epidemic of Lethargic Encephalitis in Children." *Archives of Pediatrics* 37 (1920).

———. "Lethargic Encephalitis." *Transactions of the Section on Preventive Medicine and*

Public Health of the American Medical Association, Seventieth Annual Session, Atlantic City, NJ (June 9–13, 1919).

———. "The Present Status of the Etiology of Epidemic Encephalitis." *Journal of the American Medical Association* 91 (1928).

———. and H. L. Wilcox. "Does the Virus of Influenza Cause Neurological Manifestations?" *Science* 86, no. 2229 (September 17, 1937).

Norgood, C. P. "Fundamental Problems of State Institution Farms." *State Hospital Quarterly* 9, no. 2 (February 1923).

Parry, Manon. "Sara Josephine Baker." *American Journal of Public Health* 96 (2006).

Paterson, D., and J. C. Spence. "The After-Effects of Epidemic Encephalitis in Children." *Lancet* 2 (1921).

Pearce, J. M. "Baron Constantin von Economo and Encephalitis Lethargica." *Journal of Neurology, Neurosurgery and Psychiatry* 60, no. 2 (1996).

Pilcher, Ellen. "Relation of Mental Disease to Crime, Including a Special Study of the State Hospital for the Criminal Insane at Ionia." *Journal of the American Institute of Criminal Law and Criminology* 21, no. 2 (August 1930).

Raghav, S., J. Seneviratne, P. A. McKelvie, C. Chapman, P.S. Talman, and P. A. Kempster. "Sporadic Encephalitis Lethargica." *Journal of Clinical Neurosciences* 14 (2007).

Ratiu, P., and I. F. Talos. "The Tale of Phineas Gage, Digitally Remastered." *New England Journal of Medicine* 351, no. 23 (December 2, 2004).

Ravenholt, R. T., and William H. Foege. "1918 Influenza, Encephalitis Lethargica, Parkinsonism." *Lancet* (October 16, 1982).

Reid, A. H., S. McCall, J. M. Henry, and J. K. Taubenberger. "Experimenting on the Past: The Enigma of von Economo's Encephalitis Lethargica." *Journal of Neuropathology and Experimental Neurology* 60, no. 7 (July 2001).

Rhein, J. H. W., and F. A. Ebaugh. "Affective Disorders following Acute Epidemic Encephalitis in Children." *American Journal of Psychiatry* 80 (April 1925).

Robeson, H. A. "The Children's Unit at Kings Park State Hospital." *State Hospital Quarterly* 10, no. 4 (1925).

Rosenow, E. C. "Streptococci in Relation to Etiology of Epidemic Encephalitis. Experimental Results in 81 Cases." *Journal of Infectious Disease* 34 (1924).

Sands, Irving J. "The Acute Psychiatric Type of Epidemic Encephalitis." Read to the annual meeting of the American Medical Association, Washington, DC (May 19, 1927).

Saper, C. B., T.C. Chou, T. E. Scammell. "The Sleep Switch: Hypothalamic Control of Sleep and Wakefulness." *Trends in Neurosciences* 24, no. 12 (December 2001).

Schaeffer, Morris. "William H. Park (1863–1939): His Laboratory and His Legacy." *American Journal of Public Health* 75, no. 11 (November 1985).

Sheehan, Robert. "Epidemic Encephalitis." *Psychiatric Quarterly* 2 (March 1928).

Sherman, McCall, James M. Henry, Ann H. Reid, Jeffery K. Taubenberger. "Influenza RNA Not Detected in Archival Brain Tissues from Acute Encephalitis Lethargica Cases or in Postencephalitic Parkinson Cases." *Journal of Neuropathology and Experimental Neurology* 60, no. 7 (July 2001).

Showalter, William Joseph. "New York: The Metropolis of Mankind." *National Geographic* 34, no. 1 (July 1918).

Spence, J. C. "Mental Sequelae of Encephalitis." *Lancet* 1 (1925).

Steen, Patricia. "Chronic Epidemic Encephalitis." *Psychiatric Quarterly* 5, no. 4 (December 1931).

———. "Epidemic Encephalitis." *American Journal of Nursing* 31, no. 11 (November 1931).

Stevens, Dennis L. "Streptococcal Toxic-Shock Syndrome: Spectrum of Disease, Pathogenesis, and New Concepts in Treatment." *Emerging Infectious Diseases* 1, no. 3 (July–September 1995).

Storrs, Harry. "Clinical Aspects of Encephalitis Lethargica in Children." *Psychiatric Quarterly* 2, supp. 1 (March 1928).

Strecker, E. A. "Behavior Problems in Encephalitis: A Clinical Study, Etc." *Archives of Neurology and Psychiatry* 21, no. 137 (1929).

———, and G. F. Willey. "An Analysis of Recoverable Dementia Praecox Reactions." *American Journal of Psychiatry* 3, no. 4 (April 1924).

Tilney, Frederick. "A Comparative Sensory Analysis of Helen Keller and Laura Bridgeman II: Its Bearing on Future Development of the Human Brain." *Archives of Neurology and Psychiatry* 21 (1929).

———. "With Passing Years." *Yale Literary Magazine* 24 (1895).

Triarhou, Lazaros C. "The Percipient Observations of Constantin von Economo on Encephalitis Lethargica and Sleep Disruption and Their Lasting Impact on Contemporary Sleep Research." *Brain Research Bulletin* 69 (2006).

Van Toorn, Ronald. "Encephalitis Lethargica in 5 South African Children." *European Journal of Pediatric Neurology* 13, no. 1 (January 2009).

Vilensky, J. A. "Encephalitis Lethargica." *Pediatric Neurology* 6, no. 6 (December 2006).

———. "The 'Spanish' Influenza Epidemic of 1918 and Encephalitis Lethargica." www.thesophiecamerontrust.org.uk.

———, P. Foley, and S. Gilman. "Children and Encephalitis Lethargica: A Historical Review." *Pediatric Neurology* 37, no. 2 (August 2007).

———, R. Z. Mukhamedzyanov, and S. Gilman. "Encephalitis Lethargica in the Soviet Union." *European Neurology* 60, no. 3 (2008).

Vink, Gertrude. "The Little Red School House at Kings Park State Hospital." *Psychiatric Quarterly* 13, no. 1 (March 1939).

Von Economo, Constantin. "Cruchet's Encephalomyelite and Epidemic Encephalitis Lethargica." *Lancet* (July 20, 1929).

Walters, J. "Hitler's Encephalitis: A Footnote to History." *Journal of Operational Psychology* 6 (1975).

Ward, Christopher. "Encephalitis Lethargica and the Development of Neuropsychiatry." *Psychiatric Clinicians of North America* 9, no. 2 (1986).

———. "Influence of Encephalitis Lethargica on Hysteria." *Journal of Neurology, Neuropsychiatry, and Psychiatry* 77, no. 7 (July 2006).

War Information Office. *The Medical Clinics of North America.* U.S. Army, vol. 2, no. 2 (September 1918).

White, William A. "Strange Behavior of Insane Has Meaning." *Science News-Letter* (August 17, 1929).

Witzel, August E. "Epidemic Encephalitis, Sequelae and the Psycho-neuroses." *State Hospital Quarterly* 10, no. 3 (May 1925).

NEWSPAPER ARTICLES

New York Times Archives (1915–39)

New York Evening Post Archives (1918–26)

New York Herald Archives (1918–26)

New York Tribune Archives (1918–24)

New York Herald Tribune Archives (1926–35)

London *Times* Archives (1916–30)

Altman, Lawrence K. "The Doctor's World." *New York Times*, August 24, 2007.

Anonymous. "Effects of Sleepy Sickness on Children." London *Times*, December 9, 1924.

———. "Unique Three-fold Study of Mental Ills Begun." *New York Times*, September 15, 1929.

———. "Heat Kills 3 Here; Relief is Due Today." *New York Times*, July 30, 1931.

———. "Professors of Government." *Washington Post*, April 8, 1930.

———. "Review of Frederick Tilney's *The Brain: From Ape to Man*." April 12, 1928.

———. "Waking Up to the Big Sleep." London *Independent*, July 30, 1998.

Gary, Christopher. "Streetscapes: Morrisania Hospital: A Tidy Relic of the 1920s Looking for New Use." *New York Times* (July 15, 1990).

Obituary for Smith Ely Jelliffe. *New York Times*, September 26, 1945.
Obituary for Smith Ely Jelliffe. *New York Herald Tribune*, September 26, 1945.
Obituary for William J. Matheson. *New York Times*, May 16, 1930.
Obituary for Mrs. Jane Morgan. *New York Times*, August 16, 1925.
Obituary for Mrs. Jane Morgan. *New York Herald Tribune*, August 16, 1925.
Obituary for Josephine B. Neal. *New York Times*, March 20, 1955.
Obituary for Nathan Charles Rothschild. London *Times*, October 1923.
Obituary for Frederick Tilney. *New York Times*, August 8, 1938.
Obituary for Frederick Tilney. *New York Post*, August 8, 1938.
Sacks, Oliver, and Joel A. Vilensky. "Waking to a New Flu Threat." *New York Times*, November 16, 2005.
Wright, Oliver. "His Life Passed in a Trance, but His Death May Solve Medical Mystery." London *Times*, December 14, 2002.

DOCUMENTARY FOOTAGE

"Awakenings." Discovery series, Yorkshire Television, England, 1974.
"The Crash of 1929." *The American Experience*, PBS Home Video.
"Medical Mysteries: The Forgotten Plague." BBC, England, 2004.
"New York: A Documentary Film." *The American Experience*, PBS Home Video, 2001.
"Surviving the Dust Bowl." *The American Experience*, PBS Home Video.

WEBSITES

www.americanhistory.si.edu
www.cdc.gov/about/history/ourstory.
"The Great War and Shaping of the 20th Century." www.pbs.org/greatwar.
Godl, John. "How a Wrong Can Make a Right." www.firstworldwar.com.
"History of the Neurological Institute of New York," www.cumc.columbia.edu.
"Horse Gas Masks," www.sciencemuseum.org.
"Jarrod's Story." University of Texas Health Science Center at Houston (July 10, 2007), www.uthouston.edu.
www.keyhistory.org/matheson
Kings Park Heritage Museum, www.kingsparkmuseum.com.
"Kings Park Psychiatric Center," www.lioddities.com/Asylums/KingsPark.
Macnair, Trisha. "Encephalitis Lethargica." www.bbc.co.uk.
"The Mind Traveler: An Interview with Oliver Sacks." www.fortunecity.com, July 31, 1998.

"Mystery of the Forgotten Plague." www.bbc.co.uk, July 25, 1998.

"Mystery of the Forgotten Plague." www.bbc.co.uk, July 27, 2004.

Nobel Prize nominations, www.nobelprize.org.

www.nyc-architecture.com.

www.nycsubway.org/maps/historical.

"Phineas Gage." www.brainconnection.com.

Smailes, Gary. "The Man Who Didn't Shoot Hitler." www.victoriacross.wordpress .com, March 12, 2007.

The Sophie Cameron Trust, www.thesophiecamerontrust.org.uk.

"Solving a Medical Mystery." www.action.org.uk, December 2004.

INDEX